Gender and Rural Geography

Pearson Education

We work with leading authors to develop the
strongest educational materials in geography,
bringing cutting-edge thinking and best
learning practice to a global market.

Under a range of well-known imprints, including
Prentice Hall, we craft high-quality print and
electronic publications which help readers to understand
and apply their content, whether studying or at work.

To find out more about the complete range of our
publishing, please visit us on the World Wide Web at:
www.pearsoneduc.com

Gender and Rural Geography

IDENTITY, SEXUALITY AND POWER IN THE COUNTRYSIDE

JO LITTLE

UNIVERSITY OF EXETER

An imprint of **Pearson Education**

Harlow, England · London · New York · Reading, Massachusetts · San Francisco
Toronto · Don Mills, Ontario · Sydney · Tokyo · Singapore · Hong Kong · Seoul
Taipei · Cape Town · Madrid · Mexico City · Amsterdam · Munich · Paris · Milan

For my parents, Brian and Anne Little

Pearson Education Limited
Edinburgh Gate
Harlow
Essex CM20 2JE
England

and Associated Companies throughout the world

Visit us on the World Wide Web at:
www.pearsoneduc.com

First published 2002

ISBN 0 582 38188 6

British Library Cataloguing-in-Publication Data
A catalogue record for this book is available from the British Library

10 9 8 7 6 5 4 3 2 1
06   05   04   03   02

Typeset in 10/13 pt Sabon by 35
Printed in Malaysia , LSP

Contents

Acknowledgements

Although the words in this book are mine, the process of collecting them together and putting them on paper has relied on other people in a variety of ways and I would like to express my thanks to them. I am very grateful to all those who have participated in the different research projects on which this book draws. These include policy makers, community activists and, most of all, rural women all of whom have given their time generously to answer questions and express opinions on things that often must have seemed far removed from their everyday lives. I should also like to thank the other academics who worked with me on research that has informed this book, in particular Owain Jones who was involved in the 'Rural Challenge' work and Tricia Austin with whom I collaborated on the Avon research. Some of the ideas included in the book have been 'tried out' in conference papers and I am grateful to all those who have asked questions, made comments and generally helped me to rethink my ideas; in particular Carol Morris, Jane Pollard and Rachel Woodward. At Pearson Education Matthew Smith and Louise Lakey deserve thanks for their enthusiasm, efficiency and all the other things publishers do.

Colleagues in the Geography Department at Exeter have been important in providing advice and support during the writing of this book and I am grateful in particular to Kate Brace, Mike Leyshon, James Kneale and Tony Brown. Finally, particular thanks are due to Fran Hamilton and Bill Baker for their interest in the book and for their friendship.

For some months a rumour was circulating amongst rural geographers that I was writing a book called 'Sex and Tractors' (thanks, Henry). I hope that those awaiting its publication will not be disappointed by the somewhat broader direction of the final version.

In addition the publishers would like to thank the following for permission to reproduce copyright material: Table 4.2 adapted from *Sociologica Ruralis* Vol. 34, 1994, published by Blackwell Journals (Halfacree, K.); Table 5.1 from *The Rural Economy and the British Countryside* published by Earthscan (Allanson, P. and Whitby, M. (editors), 1996); Figure 7.1 from the front cover of *Country Life* Vol. CXCV No. 11 published by IPC Media Ltd., March 15, 2001.

Whilst every effort has been made to trace the owners of copyright material, in a few cases this has proved impossible and we take this opportunity to offer our apologies to any copyright holders whose rights we may have unwittingly infringed.

1 | Introduction

In 1986 I wrote an editorial in the *Journal of Rural Studies* entitled 'Feminist perspectives in rural geography' (Little, 1986). This was an introductory paper, the purpose of which was to draw attention to the lack of published material on either gender issues or women's lives within rural geography. The paper was based on my knowledge of mainly British and North American geography and argued that the burgeoning of feminist perspectives in other parts of geography was setting rural geography apart and confirming it as one of the discipline's least progressive subject areas. I acknowledged the work that did exist (for rural geography at the time wasn't totally devoid of feminist enquiry) but stressed the need for research which not only provided detail on the patterns of gender difference and inequality in rural society but also sought to apply feminist perspectives to theoretical and conceptual debates within rural studies.

The mid-1980s was a significant period for the development of feminist approaches in geography, as I outline in Chapter 2. A concern with documenting the nature of women's lives in different places was being replaced by a recognition of the importance of gender as a factor in explaining patterns of inequality (see Bowlby *et al.*, 1989). Feminist geography was arguing the need to explore patriarchal power relations between men and women as the basis of understanding gender difference and, in particular, women's subordination. Although still rejected by some, feminist perspectives were increasingly being recognised for the major contribution they were making to geographical knowledge. It was in this broad context that my article criticised the absence of feminist approaches in rural geography and urged those working in rural studies to challenge the patriarchal nature of both the content and approach of the sub-discipline.

By not embracing the developments emanating from feminist approaches, it was argued, rural geographers were failing in two key ways. Firstly, and quite simply, they were neglecting the particular experiences of women in rural areas; gender studies had established a 'geography of women' which, while rightly criticised later for adopting a somewhat isolated and untheorised approach, demonstrated the wide range of issues which became foregrounded when the focus of geographical research shifted to women's lives. Thus, issues such as accessibility, domestic labour, community services and labour market relations, provoking excitement elsewhere, were largely ignored in the rural context. Rural geographers were failing to grasp an important opportunity to identify the patterns of women's lives in the rural community and to develop a better understanding of the manifestations of gender inequality in the countryside. Secondly, the absence of a specifically feminist approach ensured that the role of gender in broader social relations within the rural household and community was being overlooked, and as a result understanding of the community itself and the power relations within it was weakened. The lack of theoretical consideration given to gender relations within the rural community also meant that the specific contribution a rural perspective could make to debates on the form and direction of patriarchy went largely unrecorded.

During the late 1980s and early 1990s gender studies in rural geography began to make up ground. From an initial focus on farm women and their role in agriculture, studies of rural women started to filter into other areas of rural geography and feminist perspectives shifted debates from women's lives to gender relations. Mainstream issues of interest to rural geographers – labour markets (Little *et al.*, 1991), the rural community (Middleton, 1986; Stebbing, 1984), tourism (Bouquet, 1987) and the environment (Sachs, 1994) – were examined through the lens of gender and increasingly attempts were made to move away from a strict compartmentalisation of women's lives to an appreciation of the broader relevance of gender. As the volume of work on gender increased so did its status within rural geography. But although chapters in edited collections and papers in key journals began to appear, feminist studies in rural geography remained somewhat marginalised and isolated, described, even into the 1990s, by Whatmore *et al.* (1994) as a 'fugitive literature'. Only in the late 1990s has work on gender become part of 'mainstream' rural geography. *Gender and Rural Geography* is a product of this move and of the rich and varied debates that have, with growing confidence, found their way into the core of the sub-discipline.

The inspiration for this book comes partly from the history of the study of gender and rural geography but also from what remains unsaid. While a lot of ground has been covered since my 1986 article, there are still major gaps. Moreover, the work that has taken place has been somewhat fragmented and scattered. This book is, first, an attempt to bring together the various studies of gender and rural geography that have been undertaken and to draw out some key arguments from this work. Existing research has, as indicated, tended to take a somewhat fragmented approach and there is a need for ideas from this work to be collected together in a single critical account. Second, the book introduces some new material from my own research on key issues such as rural employment, community and governance. It seeks to set this material within existing debates but also to use it in the presentation of new conceptual ideas. Third, the book will include some more recent ideas from both feminist and rural geography in the examination of rural gender issues. In particular it will develop current interest surrounding gender identities, sexuality and embodiment in the context of socio-economic relations within the contemporary countryside. In so doing the book will also engage with some more recent research directions in rural geography emerging from the adoption of 'new' cultural approaches. Although aiming to bring together a diverse body of research, established and more recent, the book cannot claim to be comprehensive. There has clearly been a need to be selective in terms of both the themes addressed and the illustrative research upon which these themes draw. Hopefully, however, the book, in its review of past work, provides a substantive and detailed commentary on gender and rural geography and starts to explore some important new directions in a way that can be used as a springboard for future research.

Chapter 2 examines the development of theoretical ideas from feminist geography within which rural gender studies have been framed. It shows how the debates taking place within the discipline more widely have been applied to (and have influenced) the direction of feminist research in the rural context. The chapter traces the different phases of feminist thought from early work on gender roles through to the current focus on gender difference and gender identity and on the uncertainty surrounding the relevance of categories based on sex and gender. In establishing the background to the study of rural gender issues it is important, as well as looking at the development of feminist theory, to consider how changes in the research agenda of *rural* geography have been reflected in the scope and direction of work on gender. Since the 1986 article referred to above was written, rural societies and economies of the developed world have undergone a series of transformations, many of which impact on the issues that concern us in this book. Feminist geography has moved on but so too has rural geography; there have been shifts in the study of rural areas as well as in their character. What is of interest here is how the different dimensions of change and the different perspectives on that change have interacted. The following section outlines some of the relevant aspects of the recent transformation of rural societies and how these have come to be understood through the changing academic discourses of rural geography.

Geography, rurality and change

There have been a number of academic studies charting the development of rural geography; from the descriptive land use studies of the 1960s and 1970s, through political economic perspectives on restructuring, to post-structuralist approaches to constructions of rurality and considerations of otherness, rural geography has been subject to extensive reflection both in the UK and overseas (see, for example, Cloke, 1989, 1997; Phillips, 1998a, 1998b; Philo, 1992; Troughton, 1995). The need to reassess past directions in rural geography may be interpreted as profound interest in the histories of the sub-discipline but it also reflects the dynamic nature of the subject area itself.

Paul Cloke (1997) identifies a number of different phases in the development of rural geography and in the status of rural studies within the wider discipline of geography. He notes the centrality of the study of rural areas to the regional focus of geography in the first part of the twentieth century and the emphasis that was placed on the classification of agricultural and rural landscapes in the development of geography as a spatial science. The dominance of agriculture in the economy also ensured the prominence of rural studies. Rural communities excited little research interest, beyond the classification of settlements, and even as geographers started to become more fascinated by social scientific issues, particularly welfare and deprivation, the lives of rural people and the operation of rural society were largely overlooked. Geography from the 1960s became increasingly focused on towns and cities, encouraged by the scale and visibility of urban issues and the rapidly growing and sophisticated research agenda of urban studies (Cloke, 1997).

As authors such as Philo (1992) and Phillips (1998a) have pointed out, rural geography in the 1960s and 1970s was 'deserted of people' (Philo, 1992: 200), preoccupied by measurement and spatial classification.

Settlements were 'explained' according to location models based on agri-cultural land use and economic rationale. Social relations, seen as deriving from the organisation of agriculture and dominant economic interests, were 'naturalised', harmonious and universal (Phillips, 1998a). Even studies of migration and counterurbanisation did not focus, as Phillips notes (1998a), on the lives of rural people or on the political and con-tested nature of population change. Rather, they attempted to identify and explain migration trends in relation to certain 'universal laws' based largely on economics:

> [D]epopulation was seen to be the result of economies of scale and cumu-lative advantage of the process of centralisation, while counterurbanisa-tion was explained as the outcome of such laws as the maximisation of individual preferences, economic cost maximisation and the balancing of demand and supply. (Phillips, 1998b: 126)

There was, at this time, little interest in the development of theoretical understandings of rural space; where theory was used it was confined to the work of agricultural economists (Marsden, 1998).

During the 1980s, however, major changes occurred in the study of rural economies and societies of developed countries, changes that were inspired both by the processes of transformation affecting rural areas and by the willingness of those involved in rural research to adopt new theoretical perspectives. Rural geographers recognised the far-reaching changes associated with the processes of rural restructuring and in so doing accepted the need for new theoretical approaches to understanding the nature and influence of these processes. As Marsden notes:

> Since the early 1980s it has become increasingly clear that rural areas, both in the United Kingdom and in the advanced world, have been caught up in a much more complicated national and international political economy: a period of social and economic restructuring which has become highly diverse and fragmented. (1998: 15)

He goes on to identify the different ways in which social and economic restructuring, and its implications for rural areas, have been interpreted in the various theoretical frameworks employed by rural geographers.

Initially, research on rural restructuring focused on the economic rela-tions surrounding agriculture. Work by rural geographers sought to under-stand the background to agricultural production in developed countries and to examine, in particular, the significance of changes in the power and policy networks surrounding food production. Thus attention was directed to the vertical and horizontal linkages between different elements of the agricultural industry in an attempt to show how decision-making on the farm was related to the global patterns of investment and control (see Goodman and Redclift, 1991; Marsden *et al.*, 1993). Research focused on a range of issues including the changing nature of the labour process, agricultural land and property relations, and national and international farm policy in identifying the uneven effects of what became termed the 'farm crisis'. The end of the 'productivist era' of farming was occurring, with huge implications for the farming industry and for the economies of rural areas generally.

In identifying the nature and effects of restructuring, rural geographers began to look at the relationship between agriculture and other aspects of the rural economy and society. It had become clear that the complicated processes of change taking place were intricately related and that they both stemmed from and contributed to broader shifts in the production and consumption of rural spaces. As well as major changes in the farming industry, new demands were being placed on the countryside by a combination of industrial relocation, migration patterns, leisure practices and environmental concerns. The relationship between the factors involved in this transformation from production to consumption inevitably varied from place to place and rural geographers argued for research to identify not only the international and national patterns of change but also the more localised manifestations. They also recognised, as Marsden (1998) notes, the need for theoretical approaches to be adapted to recognise the different demands on the countryside and, in particular, the realignment of state–society relations.

> Food, the environment and the pressure for amenity are creating new and uneven demands on rural space; the accommodation of much broader consumption concerns, beyond those dealing simply with production, have begun to foster new types of rural and regional development. These new demands . . . owe their origins to a disparate array of forces both within and beyond state control and guidance. (Marsden, 1998: 16)

Marsden *et al.* (1993) suggest that to understand the 'developmental trajectories' of rural regions in developed economies we must examine four main sets of parameters. These are first, the structure of the local economy – its buoyancy and diversity; second, the demographic structure, particularly the importance of middle-class inmigration; third, politics, participation and the influence of the local state; and fourth, local cultures including attitudes to land ownership and property and ideas of community. In investigating these parameters, they suggest, it is possible to identify four broad 'types' of contemporary countryside. These they refer to as the preserved countryside, the contested countryside, the paternalist countryside and the clientist countryside.[1] This framework clearly allows the relationship between the different parameters of change to be interrogated and enables rural places to be seen as the result not of the transformation of discrete, isolated practices and policies affecting either agriculture or the community or the environment, but as the outcome of the combination of all these factors within specific locations.

This typology also demonstrates the importance of recognising the uneven nature of processes and outcomes of rural restructuring. While it stresses the interrelatedness of social and economic factors and encourages the exploration of social and cultural practices in the context of contemporary rural restructuring, it prioritises, in the detail of explanation, the economic and policy-related factors. In addition to the type of work which contributes to (and stems from) this political economic perspective on rural change, geographers have recently become influenced by the development of socio-cultural approaches within the discipline and their application to rural studies.

Phillips (1998a, 1998b) has provided a reading of the development of rural research from the perspective of social geography. He records the contribution of early rural sociologists to the understanding of the rural community and shows how geographers took up ideas on the relationship between urban and rural settlements in seeking to understand the nature and implications of population change in different localities. Phillips (1998b) notes how the development of political economy approaches in rural geography in the 1980s focused attention, for the first time, directly on social class. How class is conceptualised has subsequently been the subject of debate within contemporary rural geography – Phillips himself identifies five different approaches to a class analysis, all of which draw on Marxist political economy. Such debates have considered issues such as the changing nature of property relations in rural areas, the role of employment and capital-led social change.

The sorts of transformation taking place in the restructuring of rural areas are clearly central to the class structure of rural areas. They also

---

1 The identification and discussion of ideal types by Marsden *et al.* (1993) relate specifically to British rural areas, although none of the characteristics identified is unique to this country.

support – so it has been argued (see Cloke *et al.*, 1995) – a new under-standing of class formation; one that is not 'grounded in production' (Murdoch and Marsden, 1994). The shift from production to consumption identified as central to the process of rural restructuring has directed debate towards the influence of different class fractions in controlling access to rural space. This debate has included a recognition of the impor-tance of aspirational factors in migratory decisions and, in particular, the role of constructions of rurality in the consumption of rural spaces. Cloke *et al.* (1995) have suggested that changes to the class structure of rural areas emanating from the national and international division of labour and the replacement of a manufacturing-based with a service-based economy, have been overlain by

> social relations based on things such as skills and qualifications, consump-tion decisions and political power created through corporations and state bureaucracies. These social relations are seen to lead to new sources of social power in addition to those produced from the capital–labour rela-tion, and to the emergence of a new service class able to utilise these new sources of power. (Phillips, 1998a: 38)

Class analyses have been highly important to both the direction and output of rural geographical research. A focus on social class helped to question the early use of rurality as a causal factor in the occurrence of rural deprivation and poverty (Bradley, 1986) and served to highlight the distributive consequences of political and economic decisions. The discussion of class-based patterns also succeeded in linking the rural economy and society into more global and national processes of change (Urry, 1984). More recently, attention has been paid to the different ways in which class fractions are constituted and reconstituted, and to the power of particular elements of the class structure to influence and change the countryside (in, for example, their attitudes towards development and influence over the planning process and their role in partnerships for economic regeneration) (see Cloke *et al.*, 1995; Hoggart, 1997).

The study of restructuring and the adoption of political economic per-spectives in the 1980s allowed rural geography to throw off the criticism that it was insular and disinterested in theory. In the mid-1990s the subject was to undergo a further 'resurgence' with the application of cultural approaches, inspired in particular by a wish to 'put people back' into the rural landscape and to focus on the varying experiences of rurality. In calling for more attention to be given to the lives of those living in and visiting the countryside, rural geographers were accusing past research of taking a 'modernist' perspective. Universal laws, it was argued, had been employed to the rural economy and society in a simplistic attempt to sweep the world into 'tidy' or 'abstract' concepts and categories and to construct all-embracing theoretical accounts of rural social life (Murdoch and Pratt, 1997). Moreover, in so doing research effort had concentrated overwhelmingly on the powerful in rural society.

In 1992, in a much-quoted article, Philo called for those engaged in rural studies to direct their research energies towards an examination of what he termed 'neglected rural others'. He argued that past approaches had succeeded in 'steamrolling' the different stories of 'other' people and ignoring the experiences of a whole range of groups living in the countryside. The relative powerlessness of, for example, the poor, the elderly, women, those with disabilities, black people and ethnic minorities, served to distance them from the interests of research. Philo argued that not only were the lives of such people important to a comprehensive understanding of rural society and community, but that only by looking directly at these 'neglected others' could rural geographers hope to comprehend the nature and causes of marginalisation and inequality more generally in the countryside. As Philo has written elsewhere, his 1992 article

> urged academic rural geographers to enlarge the horizons of their studies by wondering about the worlds of many more non-hegemonic, commonly less visible, often sad and oppressed, sometimes defiant and resourceful 'rural others' than have to date been touched upon. (1997: 22)

Since Philo's original article, many researchers have added weight to his arguments in a discussion of notions of otherness and sameness in the context of rural identities and the lived experiences of rurality (see Cloke and Little, 1997; Haartsen et al., 2000; Milbourne, 1997). Work has looked not only at the value of focusing on 'the other' but also at how this might best be achieved. Authors such as Doel (1994), for example, have warned of the dangers of approaching studies of 'the other' in a way that simply applies accepted explanations, theories and language in a reaffirmation of established modes of thought and the power of 'the same'. As Philo explains, this is because

> the very equipment for thought bequeathed to scholars in the West has an inherent 'will' to translate what might initially appear as somehow 'the Other' into the comforting vocabularies of what is tried and tested as 'the Same'. (1997: 23)

Murdoch and Pratt (1993, 1997) share Philo and others' concern that rural geography has been 'partial' and research 'skewed' in its priorities, and likewise call for attention to diversity and difference within the mainstream rural research agenda. They, however, warn against a focus on 'neglected others' in a way which simply 'adds in' a concern for 'new voices' and 'new identities' to rural geography, recommending instead a more general rethinking of the core of the sub-discipline. Murdoch and Pratt (1997) argue that in attempting to situate and understand the lives of previously neglected rural identities research must consider 'how we come to know the "rural"':

> [A]t present it is unclear how difference, diversity and fragmentation can be understood, given our current frames of reference, and there is real danger that these new concerns will simply be grafted on at the margins. . . . [O]ur worry is that hidden and neglected 'Others' may remain peripheral to a sub-discipline still orientated towards the modernisation or rational organisation of the rural sphere. (1997: 55)

Despite these – very valid – concerns about the way we conceptualise marginality and otherness in contemporary rural geographical research – the dangers of a kind of research tourism that simply documents the lives of 'other' groups without challenging conventional understandings of rurality – the attention to 'neglected' rural geographies and to differing stories of rurality has served to reinvigorate the subject. Studies of previously ignored members of rural society (of black and ethnic minority people, women, new age travellers, young people and children[2]) have enriched our understanding of village life and the workings of the rural community. Such studies have provided important detail of the experiences of different groups and helped to re-evaluate conventional understandings of social relations in the countryside. The stimulus provided to the study of women's lives by the focus on rural 'others' is clearly of particular concern to this book and is further considered below.

Studies of the rural other should not be seen simply as individual attempts to illuminate the lives of particular groups but as part of the wider contribution of cultural approaches in rural geography. The cultural turn has brought with it a 'theorization of difference' (Cloke, 1997: 369) which challenges the ways in which we see rurality. Central to this work has been the notion of imagined rural spaces. In seeking to explain the differing experiences of rural people, geographers have started to look closely at the

---

2 See Little (1999) for a review of this work and for references to studies of the rural other.

cultural construction of rurality and at the myth and symbolism incorporated within past and contemporary understandings of the rural. A particular
dimension of this work has been the exploration of constructions of the
rural as an idyllic space, in which geographers have argued that dominant
representations of the rural can conceal problems and mask difference (see,
for example, Crouch, 1992; Halfacree, 1993; Jones, 1995; Short, 1991).
In such work it has been stressed that an understanding of the cultural
construction of rurality must incorporate what Jones terms 'lay discourses'
– 'people's everyday interpretations of rural places and ideas of the rural'
(Jones, 1995: 35) – together with a recognition of the power relations
through which dominant representations are produced and consumed.

The recent development of postmodern perspectives in rural geography
has, in examining the differing everyday experiences of the rural and turning attention away from explanation and analysis in favour of narrative
and description, unsettled the relationship between the rural and geographical space. As Cloke notes:

> the signs and significations of rurality have been freed from their referen-
> tial moorings in geographical space. Multiple social spaces can be seen to
> overlap what were previously recognised as rural geographical spaces,
> and the myths and symbols of rurality are recognised to pervade wider
> social spaces. (1997: 369)

Thus Halfacree (1993) speculates on the ways in which the rural as signifier
comes to challenge the rural as material space. Such speculation has led
to the coining of the term 'post-rural' (see Murdoch and Pratt, 1993) in
which the rural is seen as increasingly deterritorialised, present only in the
form of abstract signs of imagined spaces. As yet rural geographers have
refrained from delving too deeply into notions of the hyperreal, mediating
the search for virtual rurality with a continuing concern for everyday
practice (see Lawrence, 1998). There is, however, a recognition of the
importance of constructions of rurality and of the need to be sensitive to
different ways of 'seeing the rural'. There is also an acknowledgement of
the complex relations of power which determine the differing ways in
which we make sense of the rural.

Confronting gender in rural geography

The above review of the development of rural geography has made almost
no mention of either gender or women. This essentially is the task of the

present book: to add gender into the analysis of the theory and practice of rural geography, showing where the study of women's lives and a sensitivity to the importance of gender have contributed to the broader development of rural geography and to the understanding of rural spaces and society. Thus the various chapters select conventional topics of rural geographical enquiry (as outlined in the following section) and examine the relevant work that has been undertaken on gender and how it has contributed to the theoretical and empirical understanding of these topics. Chapter 2 is the exception in that it begins by looking at rural gender issues from the perspective of feminist geography. It examines how different debates and directions in feminist studies, as employed by geographers, have been embraced (or not) by rural research. It shows the specifically rural dimension to issues such as the family, motherhood, sexuality and gender identity, and suggests avenues for the application of more recent dimensions of feminist geography within the rural context.

Before discussing the content of the chapters I should like to return briefly to the development of rural geography in discussion of two areas where gender studies have made a particular contribution to more mainstream debates. These two areas are 'farm women' and 'gender and otherness'. I have selected these two focuses for special mention here because they incorporate many of the central theoretical concerns that have characterised the development of feminist approaches in rural geography. They are areas that have a significance beyond their own particular debates in rural geography. They are also areas in which much of the impetus for study and for theoretical development has come from within rural geography itself. They will be returned to at different times over the course of the next seven chapters.

The study of women in agriculture has a relatively long history (compared to other areas of work on gender and rurality), and includes a mix of theoretical and empirical research. The examination of farm women as a specific group owes much to the debates surrounding the farm family in agricultural production within developed economies. Research in the late 1970s and 1980s, first by rural sociologists but later by geographers, started to look critically at labour relations on the family farm from the perspective of women's roles and responsibilities. Early work drew attention to the range of activities undertaken on the farm by women (see, for example, Gasson, 1980) and to the importance of their contribution to the productive work of the farm (Symes and Marsden, 1983). Women's agricultural labour was seen as a function of 'lifecycle' and, as Whatmore notes (1991), of the 'compensatory relationship' that this poses between women's domestic household labour and farm work. As research developed, attention started to shift to the labour process itself, arguing that women's everyday involvement in agricultural work was shaped not simply by the organisation of domestic labour but by the broader operation of patriarchal gender relations on the farm and in the home.

The application of feminist perspectives to the study of 'farm wives' drew attention to the role of patriarchal gender relations in issues such as property rights, the ownership of capital and control of technology. Research showed how ideologies of the family and women's positions as wives and mothers underpinned the power relations on the farm. It also demonstrated how such ideologies were contested by women, especially in the context of the commoditisation of the family farm. Clearly, as Whatmore (1994) argued, an understanding of the nature of gender relations, through which the division of labour is negotiated and performed, can provide an important insight into the wider political economy of agriculture.

One example of the new insights that can be brought to research on the political economy of agricultural change by a gender perspective concerns the study of farm diversification. While conventional analyses of diversification have tended to highlight the global economic processes involved in the destabilisation of post-war agricultural regimes and related policy responses, studies of gender have focused more directly on the experiences of people living on the land and on the changes occurring in the division of labour on the farm. Gender analyses have sought to establish the changing roles of men and women resulting from shifts in the emphases on farm-based production and in so doing have drawn attention to the long-standing (and frequently unacknowledged) involvement of farm wives in activities such as bed-and-breakfast and farm tourism (see Bouquet, 1987; Evans and Ilbery, 1992). Research has also indicated broader changes to the nature of women's work resulting from diversification and to the power relations on the farm within which their labour is mediated (see Evans and Ilbery, 1996; Symes, 1991).

The study of gender relations on the farm has also included an examination of the wider food industry, including farm-based food processing and the growth of 'locally based' food provision. Women's involvement in the production of food on the farm (for example yoghurt, ice-cream, meat products, etc.) is an important area of diversification and influential in terms of the gender division of labour and women's wider involvement in the farm business. More recently, the growing demand for 'quality' food production (particularly in the UK) has increased the potential for development of farm-based enterprises in this area and as a result for women's employment. It has also been argued (see, for example, Sachs, 1991) that changes to the cultures of food consumption in developed countries, particularly the greater use of processed foods, has affected the role of the farmer's wife and the relationship between gender, domestic labour and 'productive' work. A reduction in the time spent preparing food has led, it is suggested, to an increasing propensity for women to work off the farm and for potential shifts in labour and power relations within the home.

The study of gender and agriculture has also been influenced by more recent developments in feminist geography. As explored in depth in Chapter 3, the relationship between gender identity and the rural land-scape, especially the associations between masculinity and control of the land, has started to generate interest in the context of agricultural pro-duction. Conflicting relationships have been identified between sexuality and nature; on one hand women are seen as 'closer to "mother earth" and to nature', while on the other it is men who are deemed to have the special relationship with the land that allows them to be good farmers (see Saugeres, 1998). In exploring these debates, rural geography is con-fronting recent ideas about the construction of masculine and feminine gender identities and concerns about the continuing use of dualisms in the explanation of gender relations.

More recent feminist theoretical perspectives are also articulated in the second key area of interest in the study of gender and rural geography: marginality and otherness. As noted above, women have been identified as one of those groups 'neglected' by the conventional focus of research on the powerful members of rural society, and calls for more attention to be paid to rural 'others' have provided a stimulus to studies of rural women. Such work has recovered some of the detail of women's stories of rural life as described in the forthcoming chapters of this book. While the construction of women as a 'marginalised other' has helped to shift research attention to gender differences in the experience of rural life, the need to theorise rural marginality, as discussed above, has ensured that the power relations within which these differences are reproduced and contested have not been ignored.

Research on rural women and marginalisation has also started to show sensitivity towards gender difference. There has been a growing awareness of the dangers of seeing rural women as a single category in the develop-ment of work on gender identity and on the varying levels of power and status afforded to different women within the rural community. Here work has drawn on debates surrounding the cultural construction of rurality in showing how certain gender identities are celebrated and reinforced through the circulation of dominant meanings of 'rurality'. The notion of the 'countrywoman' has been employed to represent the identity of the 'true' or 'real' rural woman – a construction which maintains considerable power in the operation of gender relations and over the daily experiences of rural women. The reinforcement of a particular set of ideas about rural gender identities in the cultural construction of the countryside in contemporary advanced societies is one of the central themes of this book and is a prominent thread in each of the different chapters.

The chapters

With the exception of Chapter 2, each of the remaining chapters of the book addresses a substantive area of rural research from a gender perspective. As mentioned above, this includes a critical review of past work on gender in that area, together with the presentation of original research material and a reflection on the direction of contemporary theoretical and empirical contributions. Dividing the ideas and material into discrete chapters has been a difficult task. Many of the issues overlap and some repetition of ideas is inevitable. I wanted each chapter to have its own internal logic, even if no one chapter can be taken in complete isolation. Some of the key ideas, such as the power of the family in the dominant construction of rurality and the strong assertion of heterosexuality, are so central to many of the subject areas that they are revisited, albeit from a different angle, on a number of occasions. Hopefully any repetition will aid understanding rather than obscure detail.

Chapter 2, as already explained, is rather different from the thematic chapters that follow. This chapter considers the development of *feminist* geography, providing a framework of mainly theoretical ideas onto which is mapped a rural perspective. The chapter presents an overview of the way particular debates in feminist geography have influenced the study of gender in the rural context. It also shows how rural geography has provided a particular take on some of the key ideas and in so doing contributed to their development and application. In general the chapter demonstrates the early reluctance of rural geographers to engage with feminist geography and to apply a gender perspective to the study of the rural economy, society and landscape. It goes on to show the more recent progress that has been made in the study of gender issues in rural geography and suggests areas in which there might be a fruitful exchange of ideas between feminist and rural studies.

Chapter 3 looks at gender and the rural landscape. The chapter provides a strong grounding in the theoretical debates surrounding the relationship between gender and nature which, it is argued, underpin responses to the rural environment. It looks at the association between femininity and nature as part of the dominance of binaries in geographical thought. The chapter goes on to show how the relationship between gender and the rural environment is played out in a variety of different ways. Particular emphasis is placed on the notion of control and on the links between masculinity and the 'taming' of the environment. The chapter shows how, despite the idea that women are somehow 'closer to nature', they are frequently seen as 'out of place' in the rural environment. This is contrasted with the use of the rural landscape to reinforce notions of masculinity in the links made between sexuality, command of the environment and male power.

In Chapter 4 I move on to examine gender relations within the rural community. This pivotal chapter argues that the rural community constitutes a central focus for the theoretical and empirical development of ideas on gender and rurality, particularly in terms of dominant constructions of rural femininity. The rural community, it is argued, promotes a construction of the 'countrywoman' which prioritises women's domestic and mothering roles and reinforces a traditional set of gender relations. The chapter explores the implications of this association between rurality, femininity and the family for both women and men, and shows how the community provides the social and physical space for the expression and contestation of conventional gender identities.

The dominance of domestic and community work in the construction of rural women's gender identity is also shown to be relevant to the gendering of paid work. Chapter 5 explores women's participation in the rural labour market, noting the problems experienced by women in accessing appropriate employment in the countryside. These problems are seen to be a function of both the characteristics of the rural labour market and the emphasis placed on women's family roles. The importance attached to motherhood as part of the rural idyll has a material effect on the characteristics of women's involvement in paid work. The constraints placed on women by both domestic responsibilities and local labour markets are detailed in Chapter 5 in a series of case studies of women's employment (and non-employment) in selected rural communities.

Chapters 4 and 5 both make reference to the power relations surrounding women's (and men's) day-to-day experiences of the rurality; power relations embedded in the assumptions and expectations associated with constructions of rural gender identities. In Chapter 6 the issue of power is looked at in a more formal context in an examination of rural governance. The chapter explores the differential access to political power of men and women in the rural community at a number of different levels. Involvement in formal channels of power is contextualised in terms of contemporary debates on rural governance and the changing cultures and mechanisms of decision-making. The chapter argues that some of the new directions and practices of governance have encouraged a masculine approach to the formation and implementation of rural policy. It also recognises, however, that the emphasis on community governance and citizen involvement may carry important implications for the gendering of power at the local level.

Chapter 7 returns to look more directly at questions of gender identities in examining rural sexuality. The focus on marginality and otherness has, as discussed above, encouraged research on previously 'neglected' groups and, as Bell and Valentine (1995b) point out, rural gays and lesbians constitute one such group.[3] Despite an acknowledgement of the need for research on the lives of lesbians and gay men living in rural areas and on the wider assumption of heterosexuality within dominant constructions of the rural environment and community, there remains a lack of published work. Consequently this chapter is probably the most speculative. It argues that it is not only important that rural geographers attempt to uncover the experiences of those living outside the mainstream rural society, but that we also develop a more detailed understanding of the relationship between sexuality and gender identity. In expanding on this point the chapter draws on some recent discussions in feminist geography on the changing and fluid nature of masculinity and femininity, showing how such discussions can inform and be informed by an understanding of the construction of gender identity in rural areas. Such discussions make reference to recent work on the body, drawing attention to an especially neglected dimension of the relationship between gender identity, sexuality and the rural.

The final chapter focuses in part on the way research is conducted, looking at the increasing use of ethnographic methods in undertaking rural research and noting the appropriateness of such methods to work on gender. As well as briefly exploring research methodologies, the chapter draws together some of the main conclusions of the book. It recognises, however, that this book only provides an introduction to many of the issues studied. It calls for research energies to be directed towards the study of gender issues in rural geography and for those working within the sub-discipline to recognise the value of a gender perspective beyond the cataloguing of difference and inequality.

---

3 Although they also recognise the dangers inherent in the classification of sexual identities in this way.

2 | Feminist theory and rural geography

Introduction

The study of gender difference and of patterns of gender inequality within rural society has been undertaken from a variety of different theoretical perspectives. A feminist perspective is not a requirement for the study of gender in any socio-spatial environment, and important work identifying differences in the lifestyles, experiences and opportunities of women and men living in rural areas has been conducted by those working outside a feminist conceptual or theoretical framework. This chapter, however, argues that a real understanding of gender difference must incorporate explanation of the underlying causes of that difference and that this, in turn, can only be achieved through the examination of gender relations. Feminist theory prioritises the study of gender relations in seeking to understand the nature of gender difference and the production and operation of gender inequality.

With the adoption of feminist theory the political nature of gender difference becomes explicit. As discussed below, the use of feminist perspectives in geography emerged in response to both the lack of consideration of gender issues and women's lives in the study of geography and also the absence of women as academics within geographical research and teaching. It was argued that the two could not be separated and that the study of gender issues in geography could only be successfully undertaken from a position which appreciated (and confronted) the under-representation of women within the discipline (Women and Geography Study Group (WGSG), 1984). The relatively late application of feminist perspectives to the study of rural geography has tended to distance rural studies from the more overt debates surrounding the importance of feminist political perspectives to the development and content of geography. It is still the case, however, that where rural geographers have chosen to adopt a feminist theoretical framework, they have contributed to a political debate that goes beyond the sub-discipline of rural studies and the specific topics included therein.

An important part of the examination of feminist theoretical perspectives within rural geography is their contextualisation within the broader scope of feminist geography. This chapter situates feminist rural geography within the development of gender studies in geography as a whole. This background is important since it helps to explain the origin of particular debates surrounding the direction of gender studies in rural geography and informs our understanding of critical changes in both the theoretical and empirical approaches to work on gender and rurality. The location of rural gender studies within a broader feminist geographical framework is also important since it helps to recognise that the real drive and inspiration for (at least) early work on gender came from *feminist* rather than *rural* geographers. Later, as will be demonstrated below, gender studies in rural geography began to be informed by theoretical developments taking place within a specifically rural research context, but this was after the initial conceptual and theoretical groundwork had been informed by feminist geography.

The chapter falls broadly into two parts. The first discusses the early articulation of gender roles by feminist geographers and the battle for inclusion in mainstream debates within the discipline. It goes on to explain the shift from description to explanation and from women to gender as emphasis was placed on power relations between men and women in the examination of gender difference and inequality. The chapter outlines these developments in general and then goes on to look specifically at how they impacted on *rural* geography. It shows how those engaged in rural research responded to the various ideas and debates emerging from feminist geography, noting the particular difficulties and excitements that a rural focus entailed. The second part of the chapter examines more recent feminist enquiry and its application within geography. Particular attention is given to notions of difference and to the current interest in gender identity. As in the first part of the chapter, the wider discussion is given a rural focus in an examination of the ways in which recent ideas have informed, and been informed by, rural geography. This discussion provides an important foundation for the issues surrounding the cultural constructions of rural masculinity and femininity that come to the fore throughout the rest of the book.

Feminist geography: early development

There are now a number of books and articles charting the emergence of feminist perspectives in British and North American geography (see Bowlby *et al.*, 1989; McDowell, 1992; McDowell and Sharp, 1997; WGSG, 1997). Some supply a view of developments as they emerged and describe the early frustrations and excitements that surrounded the gradual (and partial) recognition of gender issues in geography. Others reflect on what has taken place, situating the development of feminist geography in the discipline as a whole and identifying its contribution to wider theoretical and conceptual shifts in human geography. Some of the texts are more measured and formal analyses of the direction and content of the discipline, while others are much more personalised accounts of the importance of feminist geography to the studies and careers of individuals (WGSG, 1997). Some are entirely theoretical, while some speak through empirical case studies of people's lives. This impressive diversity is a tribute to the richness of the material that has emerged under the banner of 'feminist geography', to the struggles that have marked its production and to the challenges that it has offered to conventional geographical topics and explanations. It is not the purpose of this chapter to provide a lengthy discussion of this now substantive literature. It is, however, important to draw out some key ideas and 'defining moments' in the early development of feminist geography contained within this work, in order that we can understand where particular ideas came from and, specifically, the rootedness of important shifts in the study of feminist rural geography.

Key publications outlining the emergence of feminist geography describe the importance, in the initial stages, of 'making women visible' within the discipline. They argue that before the 1980s geographical enquiry had conventionally concentrated on topics of greater relevance to the lives and interests of men and that as a result women had been largely absent as subjects of geographical research (see, for example, Bowlby *et al.*, 1989; Bowlby, 1992; WGSG, 1984). Clearly issues such as 'the economy', 'land use' and 'transport', described as 'male orientated', are not irrelevant to women. What was being attacked, however, was the neglect of a whole set of concerns, both within and outside the traditional issues, central to the lives of the majority of women. Thus the lack of attention given to, in particular, childcare and domestic work within geographical research was seen as devaluing the contribution of women and creating a male geography in which women's activities were deemed to be neither interesting nor important. Feminist geographers responded to the male orientation of geographical subject matter by conducting research on a range of topics of immediate concern to women – topics such as patterns of domestic work (McKenzie and Rose, 1983), provision of childcare (Tivers, 1985) and mobility and access (Pickup, 1988) – and by placing women's lives at the centre of more conventional geographical interests such as urban form and retailing (see Bowlby, 1988; Foord and Lewis, 1984; McDowell, 1983). Such work not only opened up and legitimised whole new areas for geographical scholarship but also exposed the androcentric nature of traditional geography. The accepted view that there was one way of looking at the world was challenged as new research highlighted the differing experiences, opportunities and responses of women and men.

As noted in the introduction, the project of 'making women visible' in geography was a highly political one. In accordance with conventional feminist doctrine, the political nature and meaning of the personal topics filtering increasingly into geography was emphasised. Those conducting research on women's lives stressed the relevance of topics such as domestic labour and part-time employment to the wider politicised geographies of service provision, inequality and individual rights. But the introduction and acceptance of studies of the geographies of women were also fundamentally linked to the position of women as academics within geography. Women were invisible, it was argued, not only as the subjects of geographical enquiry but as the teachers and researchers of geography. A series of surveys in the late 1970s and 1980s in North America and Britain identified the gender imbalance in the staffing of university departments of geography and drew attention, in particular, to the concentration (if this word can be used to apply to such small numbers) of women in the junior ranks of the profession (see McDowell and Peake, 1990; Monk and Hanson, 1982; Rose, 1993; Zelinsky et al., 1982).[1] Women, it seemed, were making it through undergraduate and postgraduate geography courses but then mysteriously fading when it came to teaching posts and disappearing almost completely from the ranks of the senior staff.

Having begun to redress the balance of geographical subject matter and to at least raise the issue of gender inequality in terms of teaching and research in geography, feminist geographers entered what is generally seen as the second 'phase' in the development of feminist geography. This phase is often characterised by the need to move on from the study of women's lives to beginning to explain the reasons for the differences and inequalities observed. The focus of study moved from description

---

1   A more recent survey of geography departments in the USA notes some improvement in this figure but still records that only 20 per cent of tenured staff are women (unpublished).

to explanation and in so doing from women to gender. Feminist geographers agreed that it was no longer sufficient to simply identify the problems and inequalities faced by women (important though this was at the time and, indeed, continues to be) but that they needed to show why these inequalities existed. Explanations, it was argued, could not be uncovered by looking at the lives of women alone but had to situate their lives within broader patterns of gender difference. Thus, in commenting on the attention given to women and gender as part of the major 1980s research initiative within geography in the UK commonly known as 'localities studies', Bowlby *et al.* write:

> gender *roles* rather than gender *relations* has been the focus of much of this work. By *gender roles* we mean a relatively static set of assumptions about the supposed characteristics of women and men, rather than an active social process of *gender relations* through which male power over women is established and maintained. We argue that it is through gender relations that gender roles and identities are formed and reformed, and gender experienced in particular places by women and men. Thus analysis of gender *relations* not roles, and the concomitant establishment of male power, must form a key part of any explanation of the locally specific impacts and origins of economic and social change. (1986: 328) [emphasis in the original]

In turning to explanation feminist geographers were moving from a focus on women as a biological category to a concern with gender as a set of socially constructed power relations. This shift in focus was reflected in new directions and topics for research – in, for example, the study of male power in the workplace and the exploitation of women's labour in the household and the relationship between production and consumption. It also invited new theoretical debates concerning the nature of male power and its reproduction within society. Feminist geographers sought, in particular, to develop an understanding of patriarchy as the dominant form of male power and to show how it was expressed through various structures and practices within society.[2] In the course of such debates differences emerged between feminist geographers in the conceptualisation of the relationship between patriarchy and capitalism (see Foord and Gregson, 1986; Lauria and Knopp, 1985; McDowell, 1986). Some argued that although linked in social practices, patriarchy and capitalism are separate social processes which should be analysed as conceptually distinct and

---

2 See Walby (1990) for a useful discussion of six key patriarchal structures.

contingently related. Others maintained, however, that women's oppression is a function of the way in which capitalism sustains and is sustained by the reproduction of the labour force and that consequently capitalism and patriarchy are necessarily related (Bowlby *et al.*, 1989).

As well as stimulating new directions in empirical and theoretical research, the search for explanation of gender inequality led feminist geographers to look critically at the social construction of knowledge and to question some dominant theoretical assumptions that had influenced the development and practice of academic geography. In particular, feminist geography started to challenge the existence of key dichotomies that are a central plank of Western social theory (McDowell, 1992), arguing that the creation of opposites such as mind/body, culture/nature, public/private is a function of the masculine construction and control of knowledge. As McDowell notes, social theory, on which geography has drawn, has mapped these dichotomies onto gender difference in a way that associates the weaker, inferior opposite with the female. Seen as feminine they are, she argues, more 'natural' and thus excluded from theoretical investigation.

Gillian Rose (1993) writes at some length on the importance of dichotomies to the development of feminist geography. She focuses, in particular, on the nature/culture dualism, showing how the gendered assumptions it incorporates have been central to the development of geography. Rose argues that throughout the evolution of Western thought the gendered associations of the nature/culture dualism (the notion of nature as feminine and culture as masculine) have served to distance women from the creation of scientific knowledge and from the power invested within it. The control of knowledge by men is also reflected in the separation of object from subject. As Rose again maintains, masculinity defines itself through the rejection of the non-masculine and through the distancing of the masculine self from the Other. Objective or scientific knowledge similarly relies on the separation of the object from the subject.

The Women and Geography Study Group (1997) suggests two responses that have been employed by feminist geographers to disrupt the use of dualisms and to challenge the ways in which they have shaped the understanding of gender relations and women's lives. These, the Group explains, are:

> by *reclaiming* one side of a binary category – often the side which has been less valued by non-feminist geographical analyses, and by making previously invisible processes, patterns and experiences more visible and by illustrating that the boundaries drawn between binary categories are frequently more blurred than the dichotomy constructs them as, very

> often because women and men struggle to resist them. (WGSG, 1997: 115) [emphasis in the original]

As has been shown, these strategies underpin much of the early development of feminist geography and continue to influence recent and current directions.

The importance of dualisms and the social construction of knowledge are themes which are revisited later in this book – notably the relevance of the nature/culture dualism to the study of rural gender relations (see Chapter 3). Here it is simply necessary to recognise the significance of the feminist critique of the social construction of knowledge and its value in the development of feminist geography. Acknowledging the masculine assumptions behind what we study as geographers was an achievement of early feminist geography and something that the creation of a geography of women aimed to challenge. As feminist geographers began to examine more closely the underlying patterns of gender relations, the power relations that lie behind the creation of knowledge were themselves uncovered. The deconstruction of the 'naturalness' associated with women helped to challenge geography's reliance on the powerful dualisms that lie behind men's dominance within the discipline.

So far little direct mention has been made of the role of space in the development of feminist geography. Focusing on considerations of space here provides a useful way in to examining the emergence of a specific rural dimension in the study of gender and geography. As Massey and Allen (1984) argue in their book *Geography Matters!*, space and place are fundamental influences on the formulation and operation of social processes. The configuration of gender relations in any particular place is a result of the unique social, economic, cultural and political histories of that place and also of the specific local interaction of a range of global, national, regional, local and even household characteristics. Increasingly geographers have moved away from the notion of space as a mere container of social processes to see space itself as an integral part of social and economic organisation (WGSG, 1997). Thus:

> space is no longer being described as something fixed and absolute but as something that is changed by human activity. Moreover, human understandings and conceptions of space are part of the aspects of space that are being changed by this activity. (1997: 7)

Similarly the concept of place (and its use by geographers) has shifted in terms of its relationship with social processes and characteristics. Early geographers (for example Carl Ritter and Vidal de la Blache) stressed the uniqueness of place, arguing that each displayed its own set of social and economic characteristics. More recently, however, geographers have become more interested in the way different places may exhibit the same characteristics, demonstrating the global nature of many socio-economic patterns and processes and the cultural links between different parts of the world. The WGSG refers to the work of Massey (1994) to show how geographers have developed a 'progressive sense of place'. Rather than being fluid and 'defined in terms of a particular unique and distinctive location', the characteristics or distinctiveness of place is seen to lie 'in the combination of social relations juxtaposed together in place and the connections they make to elsewhere' (WGSG, 1997: 8).

The specificity of place has long occupied the minds of rural geographers. Justifying a rural perspective within the study of, for example, transport or housing has provided considerable scope for debate. The early focus on location and on the uniqueness of place directed attention towards rural areas simply as 'interesting places' – places that were, moreover, attractive to the masculine notions of exploring and taming the more remote wildernesses, central to the direction and purpose of geography in the early part of the twentieth century (see Fitzsimmonds, 1989; Rose, 1993). Cataloguing the characteristics of rural spaces and places led to prolonged debate amongst rural geographers regarding the definitions of rurality and the specific qualities or characteristics that established a place as either rural or non-rural. The need to create boundaries around places and categorise spaces led to lengthy and often circular arguments about what was meant by 'rural'. Worryingly, as common characteristics such as poor service provision and a lack of public transport were identified, the rural became imbued with causal powers and sets of 'rural problems' were established.

The adoption of political economy approaches by rural geographers in the 1980s began, however, to challenge the specificity of the rural and the view that the problems faced by rural people were *caused* by rurality itself. A consideration of the wider processes of restructuring and agricultural transformation and of, in particular, the role of capital in such processes demonstrated, as noted in Chapter 1, the global nature of many of the trends taking place in rural areas and of the implications of such trends. While for some the acknowledgement that rural places were not unique but were linked to national and international patterns of restructuring required that we 'do away with the rural', or at the very least question its validity as a defining category of explanation and research (see Hoggart, 1990), for others the legitimacy of studying the rural was preserved by a belief that while rural spaces and places were characterised by global processes, those processes were also mediated by local factors. Rurality was no longer offered as an explanatory variable but a strong argument for studying rural places as sharing certain responses and outcomes to the accumulation of capital (as well as potentially displaying unique local characteristics) ensured that the rural as a focus of research was not obliterated.

Such shifts in debates concerning the relationship between rurality and space have been significant in the development of feminist perspectives. While studies of gender and rurality have sought to draw attention to the distinctiveness of rural places, they have also located the characteristics of men and women's lives in rural communities within an understanding of wider regional, national and international gender relations. Feminist geographers have recognised the danger in seeing the experiences and circumstances of rural women simply as a function of rurality. Rather than ignoring space, however, they have stressed the interaction of social and spatial characteristics in rural gender studies.

Early feminist geography and the rural

The desire to focus on the activities, needs and problems of women came relatively late to geographies of rural areas. While urban geographers examined the daily lives of women, drew attention to the inequalities they faced in relation to the labour market, accessibility, housing and so on, and argued for the inclusion of spaces of reproduction as well as those of production within geographical study, rural geographers remained largely unconvinced. Feminist studies of rural women had taken place – indeed the debates surrounding the contribution of women to the family farm business were important in the development of the notion of petty commodity production within Marxist theory. But such studies were located firmly within sociology; they were generally not designed to draw attention to women's roles or to gender relations *per se* within the rural but rather to contribute to Marxist deliberations on the relationship between production and reproduction. While recognised as important to later studies by rural geographers on gender relations (see Little, 1987; Whatmore, 1990), they did not provide a stimulus to studies of women's roles generally within the rural community.

As noted earlier, rural geography in the 1970s and 1980s could hardly be described as at the 'cutting edge' of theoretical debate. It has long had the image, in Britain and North America, of lagging behind in terms of the key defining theoretical developments within geography, and the response to feminist geography was no exception. Notwithstanding some exciting changes taking place in the study of rural geography over the past 15 years, there has remained a strong 'head down, carry on regardless' dimension to some aspects of research. Coupled with the tendency for (parts of) rural geography to be somewhat late in shrugging off the 'explorer' image, mainstream rural geography was wary of embracing change in general but that involving feminism in particular. During the 1970s and into the 1980s, rural geographers tended to adopt a rather benign attitude to the study of gender and the advances of feminism; they reacted not so much with the hostility demonstrated in some other areas of geography, but with a rather bemused lack of interest – as if not quite comprehending what it had to do with them.

Having said this, it would be wrong to imply a total absence of rural work from the early 'add women and stir' school of feminist geography. And, indeed, studies of this ilk continue to appear today (in defiance of the notion of successive 'phases' of feminist geography, as discussed later). Research has thus sought to document the detail of rural women's lives in studies on employment (Little *et al.*, 1991), childcare (Halliday, 1997; Stone, 1990), the community (Middleton, 1986; Stebbing, 1984). In addition, work on women's roles in agriculture identified the extent of women's contribution to the farm business in documenting not only the particular tasks undertaken by women but also the centrality of these tasks to the operation of the farm business (Gasson, 1980; Symes and Marsden, 1983). As with feminist geography generally, these studies not only contributed to our understanding of the daily lives of rural women, increasing our awareness of the needs and problems that exist in rural communities, but they also helped to broaden the focus of rural geographical enquiry to include the spaces of the home and of the domestic sphere generally. Moreover, by showing how women's lives were complicated by problems of mobility and access to service provision, they started to demonstrate the interconnectivity of the public and private spheres. Much of this work is reviewed in the following chapters so will not be described in detail here.

While valuable in their own right, these studies were slow to influence the general direction of rural geography at anything like a fundamental level. They tended to be regarded in isolation as contributing another layer of information in the study of rural people and places but not affecting or challenging other parts of the subject. Slowly, studies of women's experiences began to be included in rural texts (Stebbing in

Bradley and Lowe, 1984; Little in Champion and Watkins, 1991; Middleton in Bradley et al., 1986), but even such gestures did little to shift the wider (masculine) focus of rural geography. This was a prime case of studies of women indeed being 'added in' to an existing area of geography with little real impact on other work in that area.

The development of the study of gender relations within feminist geography provided an important stimulus for gender studies in rural geography. The focus on the *explanation* of women's inequality in the study of the relationships between men and women and the operation of male power centralised the examination of gender much more within the organisation of the rural household and community. It was argued (as in other areas of feminist geography) that the experiences of women were determined at a very fundamental level by the division of labour in the family and in rural society and that this reflected the patriarchal power of men over women. Attention turned to the relationship between the domestic sphere and the 'public' world of work in analyses of the interconnectivity of production and reproduction in the countryside (see Whatmore, 1991). Such analyses included a consideration of patriarchy and its operation within the rural context.

In keeping with the spirit of political economy perspectives in rural studies, the analysis of patriarchy within the rural household, community and workplace stressed the global (and non-place specific) nature of gender relations. It also sought, however, to draw attention to what was seen as a specifically rural dimension to the way gender relations evolved and were played out within the rural family and village community. Studies, therefore, showed how women's roles could be linked to the decisions and assumptions made within the household concerning the values placed on male employment and the priorities that afforded to issues such as use of the family car, childcare, etc. (Little, 1987). Similarly, they demonstrated how underlying beliefs and assumptions surrounding the value of women's economic role and the place of women within the home underpinned their work in the community and the lack of status this afforded them. It was argued, however, that rurality itself imposed a particular influence on the operation of patriarchy – specifically that it prioritised the traditional nature of gender relations and the importance of the family and the home.

In focusing on the private sphere of the home, feminist rural geographers identified a strong relationship between the village and the family which, they argued, underpins in a very powerful way the evolution and operation of gender relations in rural society. From the late 1980s they sought, empirically and theoretically, to substantiate this argument and to apply it in the analysis of broader socio-economic relations in the country-side. While this direction was new in rural geographical studies, it drew on work conducted earlier within sociology where a tradition of research into 'community studies' (discussed in Chapter 4) had focused much more on the detail of social relations within the home. Leonore Davidoff *et al.* (1976) in particular had considered the associations between 'village' and 'home' in the nineteenth-century rural community. They argued:

> The underlying theme of 'home' was also the quest for an organic com-munity; self sufficient and sharply differentiated from the outside world. Like the village community it was seen as a living entity . . . harmoniously related parts of a mutually beneficial division of labour. (1976: 152)

They then went on to consider the importance of this relationship to women's roles and to the nature of gender relations within rural society, arguing that men were seen as the head of the 'natural hierarchy' which characterised village social relations and 'like the country squire took care of and protected [his] dependants' (Davidoff *et al.*, 1976: 152, quoted in Little, 1987).

Women's role within the home as the 'lynchpin' of the rural family was, it was suggested by feminist geographers, central to the social and economic relations of the village, both in the nineteenth century and more recently. They argued that the reality of women's lives, the in-equalities they experienced in relation to employment, childcare and other domestic duties, was rooted in a form of gender relations that depended on the symbiosis of family and community. This model, significantly, provided little scope for variation and was intolerant of change. It was also argued that the more traditional gender relations of rural commun-ities were bound up with a highly patriarchal attitude towards sexuality. Again, in keeping with the dominant image of women as mothers, they are portrayed as 'pure' and their sexuality as passively heterosexual. These ideas are taken up at greater length in future chapters. Here it is simply important to acknowledge the stereotypical views of sexuality in rural society and, in particular, the denial of aspects of women's sexual-ity and to recognise the part played by these ideas in the development of feminist analyses of rural gender relations.

As noted earlier, the evolution of feminist rural geography from the study of women's roles to the examination of gender relations brought it more firmly into the mainstream. It was recognised (not by all but by many) that the debates surrounding the relationship between produc-tion and reproduction in rural areas, for example, were significant to the understanding of the rural labour market and the use of rural services. Similarly, it became more widely accepted that gender constituted an important influence over issues such as the distribution and operation of power within rural institutions and the management of and experience of poverty (see Teather, 1994). Studies, for example, of farm businesses started to recognise that in the examination of both the day-to-day running of the farm and the wider issues of inheritance, decision-making or profitability, the nature of gender relations within the farm household was central. While it was certainly not the case that all (or the majority) of the research on rural society, economy and culture incorporated a gender dimension, there was a greater acceptance of the potential importance of a gender perspective throughout rural geography and less of a sense of partitioning off gender inequality to the limited dimension of women's roles.

More recently, inspired by ongoing developments in feminist geographies and by the ascendance of 'cultural approaches' in rural studies, work on gender within a rural context has taken another turn. Although this recent development in feminist rural geography has drawn to a greater extent than elsewhere on the debates taking place *within* rural geography for its inspiration, it is still very much linked to more general shifts in the study of gender throughout geography. Before going on to consider the nature and influence of the most recent developments in feminist geography within rural studies, therefore, we will again go back to the broader disciplinary context.

Feminist geographies and gender identity

Returning to the texts charting the development of feminist geography, a third, more recent direction or 'phase' is identified in which studies have sought to break down the universal categories of 'male' and 'female' and focus instead on the diversity and difference within genders. This phase, which we will go on to consider in more detail in this section, has revolved around the notion of gender identity, emphasising the more individualised and varied experience of gender in attempts to reveal difference and uniqueness where previously the search was for commonality. The interest in gender identity has again mirrored developments in the practice and study of feminism more broadly, both as a political movement and as an academic direction. It has also been influenced by key theoretical and conceptual debates within geography, notably those surrounding the introduction of cultural approaches into the study of geography and the development of poststructuralist perspectives.

In documenting the emerging interest in gender identity, there has been a tendency for discussions of the development of feminist geography to project a highly sequential move in which one set of theoretical arguments is replaced by another. As authors such as WGSG (1997) argue, not only is this inaccurate but it also encourages what they see as a masculine approach to the evolution of debate and knowledge. To suggest that certain perspectives or ways of seeing and making sense of the world have replaced others is to suggest the superiority of such perspectives. It also projects a very organised and united movement in which those engaged in feminist geographical research have obediently and simultaneously rejected one approach in favour of another. In reality, however, approaches can and do coexist; studies of women's roles, for example, continue to play an important part in the development of feminist geography while discussions of gender identity are not confined entirely to more recent work. Segal (1999: 16) applies a similar argument to the development of feminist theory itself in rejecting the mapping of three stages of feminist thought, asserting that 'the third stage now being labelled "post feminism" is not so removed from where many second-wave feminists came in'. In addition it must also be stressed that the various approaches articulated here are neither exclusive nor self-contained. Yet there is a sense in which it is helpful to acknowledge movement within feminist perspectives in geography; to see debates as building on what has gone before and, while not devaluing such work or the theoretical approach it adopts, recognising the developmental nature of our research. The maturation of feminist geography is critical not only to its ability to take up and work with new theoretical ideas within the discipline but also to its acceptance in parts of geography (rural geography included) that have resisted its contribution. It is also important to situate the current approaches in feminist geography in a context that acknowledges the movement within feminist thought and appreciates the sense in which we now occupy a different theoretical and conceptual 'space' – one which may not be replacing or denying other 'spaces', but is moving on from them.

New feminisms

Any attempt to situate recent feminist scholarship within the broader politics of feminism must recognise the major changes that took place in the women's movement in the Western world during the 1980s and 1990s.

There is now a substantial literature devoted to the examination of feminist politics at that time, documenting the transformation of the women's movement from a highly inclusive social movement with a belief in shared needs and common agendas, to a more fragmented and divided organisation with different and diverse needs and expectations (Faludi, 1992; Greer, 1999). Such texts generally locate the shift in the women's movement in the late 1970s and early 1980s, arguing that the emphasis on shared inequalities alienated those women who did not conform to a white, heterosexual identity. The women's movement in Britain and the USA was increasingly criticised as middle class, ignoring the particular issues faced by, for example, black, lesbian and disabled women. The 1980s, then, were marked by a decline in the solidarity that had been the hallmark of the second wave of feminism as it emerged in the 1960s and 1970s.

In Lynne Segal's recent description of the history of feminist politics this challenge to the direction of feminism is seen as inevitable. 'Any myth of women's cosy unity', she writes, 'fighting the combined subordinations of sex, race, class and heterosexism, could never hold for long' (1999: 23). Essentially the membership of the women's movement at this time failed to reflect the characteristics of the women towards whom many of its struggles were directed. As Segal goes on:

> Poverty and racism were constant preoccupations of women's liberation, both in theory and in practice (especially in the USA, where so many pioneering feminists had emerged out of the civil rights movement), but feminist groups remained overwhelmingly white and predominantly middle class. Conflict was soon breaking out everywhere, as particular groups of women expressed their sense of exclusion within the movement itself. (1999: 23)

Bondi (1993), in a discussion of gender and identity politics, argues that it was not just the composition of the women's movement that alienated certain identities. The methods used to explore gender inequality, notably consciousness-raising, by the feminist movement were also oppressive to some women. Consciousness-raising, so Bondi asserts, allows women to reinterpret their experiences within an environment that challenges the masculinity implicit in liberal humanism. The experiences that were reinterpreted, however, tended to belong to white, middle-class women and the exclusion of non-white, working-class women from these reinterpretations was felt by some to be just as alienating.

Commentators remain divided as to whether the fracturing of the women's movement heralded an end to either the political or social expression of feminism. The recognition and articulation of difference was a necessity that could not, it appeared, be accommodated within the existing women's movement as a political organisation. Yet many of the inequalities experienced by women today vary little from those of the 1960s and 1970s: the concentration of women in poorly paid and low-skill jobs, the continuation of domestic violence, the failure of many of the welfare services on which women are particularly dependent, for example. 'Old style' feminism, as encapsulated in the women's movement, however, was now seen as inappropriate in terms of contemporary concerns surrounding gender inequality. Some have argued that feminism, branded outdated, unnecessary and boring, has disintegrated as a political movement (see Segal, 1999). A more optimistic view of feminism is that it has adapted to respond to the inadequacies of the past and the new demands, and contexts, of the present.

A re-evaluation of what we mean by feminism has resulted – so it is argued (see Laurie *et al.*, 1999) – from the emergence of new femininities – epitomised by celebrated icons of contemporary culture such as the Spice Girls and representing the growth of 'girl power'. Such new femininities are more sensitive to difference, in particular in sexual and racial identities, and to the notion of multiple ways of experiencing the world. They are also more embedded within the contemporary world of symbol, representation and discourse in a way that allows us to be more aware not only of the different ways of experiencing gender inequality but of the various responses that may be made to it. Again, however, even in celebrating the value of new feminism, commentators have warned of (to borrow Linda McDowell's usage of a familiar phrase) 'throwing the baby out with the bath water'. As Laurie *et al.* (1999) maintain at the very start of their investigation into geographies and new femininities:

> Such celebrations of new femininity may be challenged by the realisation that gender inequalities remain embedded within economic and social structures. For example, women who have children still find themselves worse off than their male counterparts and women without children, and the 'feminization' of poverty continues to increase. This evidence has led some critics to argue that we need a 'new feminism' that will address such inequalities while capitalising on the new female confidence that exists in Britain. (1999: 1)

What is of interest here is the relationship between the kinds of changes taking place in feminism as a social and political movement and those embraced within feminist scholarship. As noted at the beginning of this section, feminist approaches have increasingly focused on the notion of difference, not only in terms of understanding the actual experiences of women but also in challenging the culturally specific assumptions surrounding the relationship between masculinity and femininity. This shift is summed up neatly by Barrett and Phillips (1992):

> Feminists have moved from grand theory to local studies, from cross-cultural analyses of patriarchy to the complex and historical interplay of sex, race and class, from notions of . . . the interests of women towards the instability of female identity and the active creation and recreation of women's needs or concerns. Part of what drops out in these movements is the assumption of a pre-given hierarchy of causation waiting only to be uncovered. (1992: 6–7)

The move away from grand theory in the emphasis on individual and localised difference has led to some very important work on the characteristics of women's lives, encouraging a growth in studies of groups of women previously neglected by research. In geography this has been evident in work on disabled women (Butler and Bowlby, 1997), black and Asian women (Dwyer, 1999) and young women (Leonard, 1998). It has also prompted a re-evaluation of the direction of work on gender, showing how feminist scholarship has tended to apply culturally specific, Western notions concerning the social construction of male and female categories to the study of gender more generally (McDowell, 1991). The focus on difference and diversity has also, however, led to what McDowell (1992) has identified as a new scepticism about the use of gender as analytical category. There has been uncertainty surrounding the validity of the very categories 'woman' and 'man' and of the use of universal concepts such as patriarchy. These concerns have been shared by geographers. As Laurie *et al.* (1999) note, while geographers have come to engage with the notion of gender as fractured and fracturing they have also provoked debates about the 'stability and utility of gender' as an analytical tool.

McDowell (1992) believes, however, that the 'profound pessimism about the future of gender as an analytical category' is unnecessary. Gender scepticism should not mean that we deny or devalue past theoretical or empirical work but that we appreciate that such work was essential to the emergence of current excitements in feminist scholarship. As Bordo (1990) (quoted by McDowell) writes:

namely sexuality and the body. Both have attracted a vast literature in recent years in feminist studies generally and also within feminist geography. Here there is space only for a limited consideration of the role of each in the conceptualisation and performance of identity, but throughout the book we shall return to these key issues in the discussion of gender identity in rural areas. The scepticism surrounding the validity of gender as an analytical tool (referred to earlier) fuelled debate on the separation of the categories 'sexuality' and 'gender' and on the relationship between biological difference, gender identity and sex. Increasing attention was placed, within this debate, on the use and control of the body in the performance of sexuality and, in particular, on the power relations incorporated within bodily expressions of sexuality.

The starting-point for debate is the argument that the power relations of sexuality cannot be reduced to those of gender (Segal, 1999). Feminists have long debated (and disagreed about) the relationship between sex and gender. As Pringle (1999: 249) notes, throughout most of the twentieth century sex has been seen as a term that linked sexuality and gender and established the characteristics of being a 'real' man or woman (i.e. engaging with 'normative heterosexual relations'). More recent theorising, particularly by feminists, has suggested that the relationship between the biological, sexed body and gender identity is more complicated and that 'our biological and social selves are mutually constituted' (Laurie et al., 1999: 4). Focusing on sexuality allows for a more fractured and ambiguous sense of identity, and the idea that there are many different ways of 'being a woman' or performing that gender.

Queer theory has contributed significantly to feminist debates on sexuality and gender, growing rapidly in popularity in the 1990s in the context of poststructuralist work and a focus on the body. Queer theory challenges what it sees as the 'exclusionary and strictly gendered identities of "lesbian" and "gay"' (Segal, 1999: 56). Emphasis is placed on the deconstruction of identity in a celebration of the lived experience of a whole range of sexual acts. As Butler (1990) argues, through such dissident sexual acts, queer theorists and activists represented 'gender trouble' in the disruption of the 'heterosexual matrix' within which sex and gender binaries are secured. In discussing the use of drag to disrupt sexual and gender identities, for example, Bell et al. (1994) write:

> Drag can...successfully parody the seriousness and essentialism of heterosexuality. Indeed, the mimicry of heterosexuality by gay men and lesbians has the potential to transform radically the stability of masculinity and femininity, undermining its claim to originality and naturalness.

> Heterosexuality is a performance, just as constructed as homosexuality, but it is often presumed and privileged as the original. (1994: 33)

Geographers have worked with debates on sexuality in the production of an important literature on the relationship between sexuality and space. This work has looked at both the experiences of different forms of sexuality within space and the role of space in the performance of sexual identity (see Bell *et al.*, 1994; Bell and Valentine, 1995a). It has drawn attention to the 'mutual relationship' between space and identity in the examination of what have become termed gay and lesbian spaces (see Adler and Brenner, 1992; Knopp, 1990; Valentine, 1993). In pointing out that it is acceptable to be homosexual in certain spaces but that such spaces are then seen as having been disrupted or 'colonised' by gays and lesbians, Bell *et al.* (1994) argue that space is presumed to be heterosexual. They argue that geographers must recognise this fundamental assumption in their work and in doing so must question why this is the case.

> We need to understand the straightness of our streets as an artefact; to interrogate the presumed *authentic* heterosexual nature of everyday spaces. (Bell *et al.*, 1994: 32; emphasis in the original)

These authors go on to examine a number of different homosexual identities, the skinhead, the butch dyke and the lipstick lesbian, in the playing out of different sexualities in space and the potential of such transgressive use of space for the disruption of heterosexual identities.

The focus on sexuality, and in particular on its performance, by feminists has fuelled a growing interest in the role of the body in the construction of identity. The treatment of the body by feminists has varied over the course of the twentieth century. The early feminist movement sought to play down the characteristics of male and female bodies; women were essentially seen as imprisoned by their bodies – bodies were thus problematical and something that needed to be transcended (Pringle, 1999). Later, feminists sought to draw attention to women's bodies as things to be celebrated and valued, especially their reproductive capabilities. Both interpretations, however, have been criticised as reproducing the mind/body dualism in which the body is seen as negative and in control of the mind. Women's bodies, in particular, were viewed as unpredictable and irrational and thus in greatest need of restraint.

More recently, feminists have come to see the body not as a passive object but rather as an active part of our identity. The particular characteristics of the male and female body have been regarded not so much as a material or anatomical entity but rather as a product of contemporary society – of expectations, meanings and values that are culturally produced. Here again, the view of the female body is often negative, as 'other' to the more competent and predictable male body. Feminist theorists working on the body have drawn heavily on the work of Foucault in arguing against the idea of the natural body in favour of a view of the body which sees it as essentially a surface of inscription, 'significant mainly in terms of the social systems or discourses that construct it' (Longhurst, 1997: 488). Foucault's work is particularly useful in conceptualising the body as a site of power. He used the history of sexuality and the discourses employed to regulate and control the body to show how it was implicated in the construction and operation of regimes of power. Foucault's work has been influential in subsequent feminist arguments surrounding the use of women's bodies in the media, and their control through exercise, diet and dress, to reinforce hegemonic constructions of femininity (see Valentine, 1999).

For geographers the body has become an important focus of research as a location in itself, as 'the geography closest in' (Rich, 1986: 212). A substantial literature has grown up within geography looking at the relationship between the body and gender identity (see, for example, Longhurst, 1997, 2000; McDowell, 1999). While the physical appearance of the body, the expectations imposed by society on the size, shape and presentation of the body and also the debates surrounding the body and disability have for some time been of interest to feminist geographers, the body is increasingly being recognised as relevant to a whole range of other areas of geography. Thus geographers working on the labour market and changing structures of employment, for example, have started to look at the changing demands being placed on the appearance of employees in many service industries, and at the influence of and implications for attitudes towards body shape and size (Leidner, 1991; McDowell and Court, 1994). The requirement for women working in the service sector, for example, to conform to certain expectations in the presentation of their bodies (in the ability, for example, to fit into particular uniform sizes) is seen as indicative of the continued operation of patriarchy in the workplace. Geographers interested in leisure have also started to focus on the body, examining the involvement of both men and women in certain sporting activities (e.g. aerobics and body building – see Johnston, 1996; Mansfield and Maguire, 1999), arguing that such involvement is frequently (and sometimes entirely) motivated by the wish to conform to dominant bodily representations of masculinity and femininity.

The growth in interest in consumption within geography has also fuelled (and been fuelled by) studies of the body. Work has looked at how the consumption of things like food (Valentine, 1999), clothing (Crang *et al.*, 2000) and perfume (Little and Pollard, 2000) has been undertaken within a set of understandings about the presentation of the body and of the value ascribed to particular bodily forms and appearances. As Longhurst writes, the body may be seen 'as a site of cultural consumption, a surface to be etched, inscribed and written on' (1997: 488). In some cases studies have also looked at the use of the body as a way of disrupting or destabilising hegemonic gender identities. Thus, Johnston (1996) has argued that while some women may visit the gym and take part in fitness and body-building regimes in order to attain the required body shape, others may exercise in an extreme way to allow them to develop bodily appearances that challenge accepted views of how women's bodies 'ought' to look. In aspects of consumption such as piercing or tattooing, the aim of some is to resist conventional body images and to disrupt the dominant relationship between the body and gender identity.

Social geographers and geographers interested in health have also drawn attention to the body in relation to constructions of and attitudes towards disability. They have shown how disabled people may be constrained in their use of space (Butler, 1999; Imrie, 1996) by the physical design of the built environment. Work has also focused on the social construction of disability and stressed the importance of dominant attitudes towards the notion of the 'fit' body in Western society in terms of the experiences of disabled people. What is clear in this range of examples is that the body is an important location for the grounding of our personal identities. It is equally clear that bodies are not fixed or static but are discursively produced, and there is a geography to that production. In focusing on the body geographers are again challenging the dualisms between, for example, the self and the other, the mind and the body, masculine and feminine. As Longhurst asserts:

> any upheaval of the dominant/subordinate structure between mind and body or between gender and sex, will threaten the privileged term's unquestioned a priori dominance in the discipline. (1997: 495)

The final part of this chapter will now go on to consider how the concept of gender identity, together with the theoretical concerns highlighted in the debates surrounding the construction and performance of identities, has informed feminist rural geography. While the ideas talked about above are relatively new to rural studies (and hence the section will not be reporting vast quantities of original work), it is worth examining the emerging and potential importance of this perspective to the study of gender and rurality.

Gender identity and rural geography

As noted above, the 'cultural turn' in rural geography has provided something of an impetus for the study of rural gender. The plea to look more closely at the lives of 'neglected others' as part of this cultural turn has, as seen in Chapter 1, helped to focus mainstream academic attention on the experiences of rural women. Moreover, the emphasis on difference and diversity in the study of identity has encouraged rural geographers to think beyond the categories 'women' and 'men' in an examination of a range of different sexual and gendered identities. While such work has been met with excitement and enthusiasm, it is still in its relative infancy.

The examination of rural gender identities by geographers has been strongly influenced by debates on the cultural construction and representation of rurality. It has been argued, for example, that the rural idyll as the dominant construction of rurality in Britain and other Western capitalist countries encompasses strong expectations of the identities of men and women and of the gender relations that exist within the home and the rural community. The details of these expectations are discussed in the other chapters of this book. Here it is simply important to recognise the key argument of the debate, namely that while gender identities in rural areas are multiple and fluid, there is a set of characteristics associated with the rural woman and (arguably less so) the rural man through which their gender identities are defined. These characteristics place emphasis on the conventional family roles of women and men and on the economic and social relations which support them. In addition, dominant constructions of rurality also assume a conventional sexuality; rural identities are heterosexual.

In discussion of these dominant characteristics of gender identity it has been argued that those who do not conform are marginalised in terms of a broader rural identity (Hughes, 1997a). To belong to the rural community, then, is to comply with accepted notions of being a woman or a man, and a failure to do so threatens that sense of acceptance within rural society. Given the dynamic nature of identity, however, it is also argued that any relationship between gender and acceptance may be fluid and may change over time. Specifically, it has been suggested that women may perform different aspects of their identities in particular situations such that marginality may relate more to certain gendered identities than to particular people (see Little, 1997a, 1997b).

The focus on gender identity in the rural community has also encouraged the development of work on sexuality. Just as feminist geography generally has drawn attention to the importance of sex in the construction of identity, so the application of feminist perspectives to the rural has sought to do likewise. Again, as is discussed in detail in the following chapters, links have been made between different constructions of masculinity and femininity and rurality. In making these links, work has drawn on debates surrounding the relationship between nature, landscape and sexuality (see Saugeres, forthcoming; Woodward, 1998). A number of studies have looked at the performance of sexuality in rural communities (see Bell and Valentine, 1995a; Fellows, 1996; Kramer, 1995), again using the idea of marginalisation in discussion of the strong dominance of heterosexual sexuality in the countryside.

The examination of sexuality in rural communities has started to generate interest amongst rural geographers in the importance of the body in sexual and gender identities. As yet there has been very little direct work on the 'rural' body, although the implications of hegemonic constructions of rural masculinity and femininity for the way in which bodies are perceived, presented and modified are starting to be included in work on rural gender issues (see Woodward, 1998; *Rural Sociology*, 2000). There is scope in these debates for thinking about a whole host of issues including not only the role of attitudes towards the body in the construction of rural gender identities, but also how such attitudes may influence issues such as women and men's use of space within the rural community, the gendered nature of rural employment (see Bryant, 1999), or, more obscurely, the consumption of food and fashion (Agg and Phillips, 1998).

These debates clearly call into question, again, the agency of space and, in particular, the integrity of the space which we refer to as rural. Again, as with the study of rural gender relations, those working in this area have argued not only that the rural is worth studying in that it adds richness to our understanding of the relationship between space, place and gender identity, but that there is also a particular association between *rurality* and gender identity that goes beyond the specificity of individual places; that is, there is a shared understanding within rural communities of gender identity. This understanding is not fixed or uncontested but it exists as a very strong influence on the way gender identities are constructed, perceived and practised. While not uniquely rural, the expectations surrounding gender identities are implied in rural areas in a way that is part of the social and cultural relations of the countryside. Responses to attempts to contest dominant gender identities will vary from place to place but what is argued here (and throughout this book) is that a set of shared and accepted meanings about the nature of gender identities continues to circulate and to inform (and be informed by) both our day-to-day experience of and wider responses to the countryside.

So, although work on rural gender identities has only recently begun to emerge, this particular theoretical development in feminist geography clearly provides considerable scope for the future. The consideration of hybridity and fluidity in the context of gender identity allows a more sophisticated examination of rural marginality and helps to establish more firmly the links between the social and cultural construction of rurality and the particular aspects of gender roles and gender relations that are valued (within dominant constructions). Focusing on identity provides an opportunity to consider different aspects of the ways in which individuals 'do gender' – in particular, key elements of identity such as sexuality or ethnicity – and the ways in which gender is contested. Aspects of this contestation may take a particular form in the countryside, as feminist rural geography is beginning to appreciate.

Conclusion

In this chapter I have attempted to provide an overview of some key theoretical debates in feminist geography and to show how such debates have informed, and been informed by, developments in the study of gender in rural geography. The discussion of these different theoretical perspectives has taken a somewhat sequential form although, as pointed out in the main body of the chapter, this is not supposed to imply the replacement of one set of ideas by another or the exclusivity of the different strands of feminist thinking. I have, in the chapter, tried not only to link gender studies in rural geography to the broader development of feminist geography but also to situate feminist geography within a series of debates in feminist studies. In so doing I have shown how the political context has remained central to the progress of academic feminism and how, in particular, the various debates, uncertainties and changes within the women's movement in North America and the UK have been reflected in the different academic 'turns'. Some of these links seem far removed from either rural studies or the reality of people's lives in rural communities. They do, however, underpin many of the ideas that run through this book and that help to situate the local within a broader theoretical context. Hopefully this chapter has helped to demonstrate, at least in part, these theoretical connections.

The chapter has shown, then, how gender studies in rural geography have been informed by feminist work on gender roles, gender relations and, more recently, gender identities. In the chapters that follow the detail of research and publications in these areas will be examined. Much of the discussion of existing work will draw on the first two of these, studies of gender roles and gender relations. In developing ideas in a more contemporary context and thinking about the future of work on gender, greater emphasis is placed on the notion of gender identity. As has been argued above, the focus on gender identity has provided scope for the consideration of a much wider range of theoretical and empirical issues in the study of feminist rural geography. It has encouraged, in particular, work on femininity and masculinity and stressed that the study of gender must go beyond the broad categories 'women' and 'men' in the examination of the shifting and multiple nature of the identities through which these binaries are represented and performed.

This focus on gender identity and on the socio-cultural construction of masculinity and femininity is undertaken here not only in recognition of the potential value of its contribution to feminist rural geography but also with some hesitancy regarding the possible loss that such a perspective can entail for the political basis of feminist geography. The study of gender identities has been both congratulated and challenged for its celebration of difference. As noted above, some theorists have been criticised for allowing the attention to difference to cloud the basic identification and understanding of gender inequality. Diane Coole sums up this tension very succinctly:

> Our own studies have increasingly shown sexual inequality to be extremely complex and diffuse, as well as revealing gender identity as a diverse and ambiguous phenomena. In one sense we can use this knowledge politically to deconstruct binary notions of sexual difference. But at the same time our more sophisticated theorising has tended to dilute a formerly more incisive representation of opposition and oppression, which makes collective action difficult in the political and economic domains. (1997: 19)

We have been seduced by discourses that distrust structural analysis and emphasise fragmentation. Coole calls this a retreat from engagement with the real and continues:

> Of course for us post post-structuralists the cultural and discursive structures that construct gender identities remain an important site of analysis and contestation. But we also need to consider afresh the roles and deprivations that are imposed by a system which does identify us as female subjects inhabiting women's bodies. (1997: 21)

It is important that we recognise and incorporate the potential of studies of gender identities in our work on rural geographies. It is also crucial that we do not allow such work to obscure the identification of some of the basic and continuing inequalities that affect women in the countryside. This chapter has gone some way towards establishing the variety of concepts and approaches that can be used to inform our understanding of gender and of rural women's experience of inequality. The next chapters will apply these to the study of different aspects of rurality at a range of scales. In so doing they will make use of a rich and varied body of work but they will also attempt to contribute to that work themselves in applying new perspectives to the study of gender in rural areas and in the recovery of the spaces of rural life.

3 | Gender, nature and the rural landscape

Introduction

Chapter 2 having established the broad theoretical development of feminist rural geography, the following chapters turn to more specific topics within rural gender studies. The chapters all review past material of relevance to a feminist perspective within these areas and examine current debates of particular concern and interest. Each of the chapters draws on material from across the different directions that have influenced feminist theory as applied by rural geographers and discussed in Chapter 2. As noted, however, the emphasis is placed on the more recent theoretical developments and their application to the particular dimensions of rural landscape and society identified.

This chapter focuses on the rural landscape and environment and on the rather broad and complex topic of nature. The intention is to look at the ways in which the rural landscape itself can be seen as gendered – that is, reflecting and sustaining particular conceptualisations and social and cultural constructions of femininity and masculinity. The chapter also examines the gendered nature of responses to the rural environment and shows how such responses are critical to our everyday experiences of the countryside in all sorts of important ways. The relationship between landscape and gender, it is argued, influences our working lives, our recreation, our use of space and our consumption practices. Our gender identities are formed in response to these influences, embracing but at the same time shaping the complicated associations between people and the landscape.

The chapter draws heavily on the theoretical ideas introduced in Chapter 2, not only in terms of reviewing past work but also in thinking about current ideas in the relationship between gender and rural landscapes.

Work on the body, in particular, is central to many of the more recent debates on how representations of rural landscapes incorporate and reflect strong assumptions regarding masculinity and femininity. The chapter begins with a more theoretical discussion which focuses on key questions such as the relationship between gender and nature, and the construction of gender identities in landscape representation. In the second part of the chapter these ideas are taken up in an examination of the more practical implications of the gendered nature of rural landscapes, looking in particular at the ways in which gender identity is incorporated in aspects of employment and leisure in rural areas and in people's use of space within the countryside.

Gender, rurality and the social construction of nature

Recent years have witnessed, as Whatmore (1999: 23) acknowledges, a 'reawakening' of debates within critical human geography on the 'question of nature'. Such debates, inspired by a range of intellectual impulses and political uncertainties,[1] have been both wide-ranging and stimulating. They have caused geographers to look again at the relationship between the 'natural' world and the 'social' world and to question ways in which the nature/society binary has continued to inform our modes of explanation. Authors (such as Castree, 1995; MacNaughten and Urry, 1998; Wilson, 1991) have discussed various aspects of the complex and shifting interplay between science and nature. Such work has challenged the 'natural realist' perspective that portrays nature as consisting of 'substantive entities', the bedrock of the 'real' world. It has also voiced some criticism of the social constructionist work in which nature is seen as an artefact of the social imagination (Whatmore, 1999).

The intricate and complex debate surrounding what actually constitutes 'nature' is beyond the scope of this book. Here I am concerned mainly with the ways in which the relationship between nature and gender have been played out in our understanding and treatment of the rural environment and landscape. Having said this, it is clear that shifting ideas about the links between nature and society (and what nature is) are highly important to gendered meanings of environment/landscape and to the embodied practices through which they are reproduced.

---

1 Debates that have been given particular resonance by the current concerns over genetically modified organisms (GMOs) and the 'naturalness' of certain agricultural practices, including cloning.

Recent discussions surrounding the meaning of nature have destabilised the automatic association between nature and the countryside. Work in the 1980s and early 1990s (see, for example, Goodman *et al.*, 1987; Goodman and Redclift, 1991; Marsden and Little, 1990) on the increasing importance of technology to agriculture production began to question the 'naturalness' of the rural environment. The transformation of aspects of the farmed landscape, the loss of flora and fauna and the lack of controls on farm pollution (see Lowe *et al.*, 1986, 1997; Pye-Smith and Rose, 1984) raised concerns that the countryside was becoming a factory floor for the production of food and that 'science' was replacing 'nature' as the key force in farming. More recently, growing fears about the safety of food and, in particular, the use of GMOs have helped to further unsettle the belief in the relationship between nature and the countryside. At the same time interest in urban wildlife and in zoos has started to raise questions about the reassertion of nature within the city (see Anderson, 1997).

While there are clearly grounds for rejecting the automatic association between 'nature' and 'country', there is still a sense in which the countryside is, if not entirely natural, more natural than the city. Popular understandings of rurality, though mindful of the ambiguities imposed by contemporary agricultural practices and other human impacts on the landscape, still uphold the countryside as 'closer to nature' than the town. This is a relationship which, as Michael Bell (1994) describes in his entertaining ethnographic study of an English village, underpins many responses to the rural landscape and community. Bell argues that the belief that the rural environment is more natural than the urban is central to the social and cultural construction of contemporary rurality and is evident in attitudes towards the rural landscape and in ideas of rural community and society. He suggests that the views of the villagers in his research (the Childerleyans) in maintaining the idea of rural dwellers as 'closer to nature' reflect a pastoralism firmly located within idyllic constructions of rurality. He sums this up as follows:

> Villagers vary in what they consider nature to be. Some connect it with a spirit they call God, others with God-like ideas they call 'balance', 'goodness' and 'purpose' . . . But virtually all the villagers use the framework of pastoralism – the idea of the gradient between ways of living close to nature and ways far from nature – to apply nature to life. Through pastoralism they connect the country and country people with nature and the natural. (Bell, 1994: 136)

These ideas surrounding the pastoral and its centrality to the notion of the rural idyll are taken up again in the next chapter in a more detailed discussion of the meaning of the rural community.

As briefly noted in Chapter 2, the separation of city and country, so central to the dominant constructions of rurality, is closely linked to the nature/culture dualism within Western thought. Indeed, as Cronon has argued, the 'long dialogue between the place we call city and the place we call country is the key to understanding the relation between society and nature' (1995: 54; quoted in Larsen, 1994). Similarly, Fitzsimmonds (1989) contends that the distinction between nature and culture became 'especially meaningful as part of the emergence of a division between the rural and the urban during the development of industrial capitalism' (Rose, 1993: 73). This division, Fitzsimmonds argues, is responsible, at least in part, for the failure of radical geographers to get to grips with the theoretical issue of nature. Nature is detached from the city and thus from intellectual life and modes of explanation.

> The pervasive view that Nature is external and primordial is uncon-
> sciously confirmed by our placement as intellectuals in a spatially organ-
> ised society in which 'intellectual work' and 'intellectual life' are urban.
> (Fitzsimmonds, 1989: 107)

Rose (1993) argues that the binary oppositions, nature–culture, country–city, have been aligned to other dualisms within human geographical thought and that the terms on either side of these oppositions have been conflated. This, she suggests, is 'encouraged by the implicit gendering of oppositional terms: the terms are engendered through the nature/culture opposition' (1993: 74). There is a long history of feminist thought that has explored the basis of the nature/culture distinction and discussed its wider importance for gendered power relations and the subordination of women (as indicated in Chapter 2). While these discussions are of relevance here, the central concern is with the ways in which the gendering of the nature/culture binary, the association of nature with female and culture with male, has been reflected in the representation and interpretation of the landscape and rural environment.

Gender and landscape

As many geographers have pointed out, how we see and make sense of the landscape reflects one of the discipline's 'most enduring interests'

(Rose, 1993: 86): the relationship between nature and culture. The study of the landscape by geographers has, as we might expect, a long history. This history has been documented in detail by other authors (see, for example, Daniels, 1993; Meinig, 1979; Rose, 1993) from the emergence of the term *Landschaft* in nineteenth-century Germany through to more contemporary ideas surrounding the symbolic meanings of landscape in the work of cultural geographers. Early writing established the study of landscape as a form of spatial knowledge, in which the landscape could be understood according to scientific and systematic theories which governed the interaction between the environment and society.

> Landscape then was never a self-evident object in geography. A theoretical framework always structured its interpretation; it was an analytic concept which afforded objective understanding. (Rose, 1996: 342)

In the early twentieth century geographers such as Carl Sauer began to reject the environmental determinism implicit in these classifications based on morphological type in the development of an approach which saw landscape as a product of the dynamic relationship between a culture and its environment. However, such an approach still suggested a scientific process whereby the natural landscape became transformed into a cultural landscape; a process, moreover, which still assumed nature and culture to be separate realms (Till, 1999).

Interest in landscape dwindled in British and North American geography of the 1950s and 1960s as 'spatial science' came to dominate and landscape studies were dismissed as overly subjective (Cosgrove, 1985). A brief revival in landscape studies amongst British geographers was stimulated by the 'perception studies' of the 1970s and by work on environmental evaluation but it was not until more recently that a sustained reawakening of interest in landscape within geographical study occurred. This more recent interest reflects a recognition of the meanings associated with landscape and of the power relations incorporated within its construction. Cultural geographers have extended this work in discussions of 'ways of seeing', arguing that the meaning of landscape is not simply about the interaction between society and the environment but is itself a 'gaze' through which we make sense of the relationship between society and the land. As Rose succinctly puts it:

> geographers have stressed the construction of the look of the landscape and have argued that landscape is a way of seeing which we learn . . . a landscape's meanings draw on the cultural codes of the society for which it was made. These codes are embedded in social power structures. (1996: 344)

With this recognition of the importance of ways of seeing has come an interest in different forms of landscape representation and in ideas of reading landscape as text.

The ways in which power relations can be seen to be embedded within the landscape have been illustrated by a number of geographers (see, for example, Cosgrove and Daniels, 1988; Matless, 1995). They have looked, for example, at the representation of eighteenth- and nineteenth-century landscapes of agricultural production, arguing that such landscapes (and the ways in which they are depicted in art and literature) reflect the social, cultural and political authority of the landowner over the agricultural worker. Other examples include Daniels' (1999) discussion of the power of early industrialists as depicted in the painting of cotton mills in the rural landscape. He describes how such paintings reminded the observer of battleships, 'the only structures to rival them in size and power' (Daniels, 1999: 58).

One of the most frequently cited examples of the depiction of power relations within the production and consumption of the landscape is Thomas Gainsborough's eighteenth-century painting of Mr and Mrs Andrews (see Rose, 1993). Studies have drawn attention to various insights into capitalist property relations that can be read from the landscape; the proprietary attitude of the couple marks them out as landowners while the absence of workers and the obvious signs of agricultural 'improvement' signify their class position and social status.

As feminist geographers have pointed out (see Monk, 1992; Nash, 1994, 1996; Rose, 1993) there are other power relations that can be read from representations of the rural landscape, namely those of gender. Our ways of seeing the landscape, they argue, embrace very powerful ideas about not only the position of women and the relationship between men and women within the rural environment, but also the association between femininity and nature. Thus, as Rose (1993) argues, the painting of Mr and Mrs Andrews can also be read as a comment on the differing relationships of men and women to the land. Mr Andrews, standing as he does in the picture, gun in hand, ready to stride off to go shooting, has a very different relationship to the landscape around him from that of his wife who sits static beneath a tree. Mrs Andrews, Rose argues, is part of nature, the tree under which she sits symbolising her family and her natural role as mother.

> Landscape painting then involves not only class relations but also gender relations. Mr. Andrews is represented as the owner of the land, while Mrs. Andrews is painted almost as a part of that still and exquisite landscape:

the tree and its roots bracketing her on one side and the metal branches of her seat on the other. (Rose, 1993: 93)

The uncovering of gender relations in the representation of the landscape is a complex business because, as noted above, the 'ways of seeing' are also gendered. The male gaze imposes a particular interpretation on the relationship between gender and the landscape which needs to be appreciated in looking at the conclusions that are arrived at about, in particular, the feminisation of nature in landscape representation.

It is clear that landscape representations contain strong depictions of gender relations within the countryside and that much can be learnt about the role and status of men and women and of the relationships between them from their inclusion in the landscape. Landscape painting reflects powerful images of the rural idyll and of the harmonious – and highly gendered – relationships within the rural community. One important aspect of these gendered images within the landscape is the way in which they encompass the relationship between women and nature. Here I show briefly how ways of seeing the rural landscape have reinforced the idea of nature as female and in so doing have contributed to the assertion of masculinity and of male power over women.

Rose (1996) uses European and North American paintings of the nineteenth century to illustrate her argument that landscape art has feminised nature. She talks of the ways in which images of women as beautiful, sexual and mysterious were fused with those of lush, green and fertile landscapes. Women were also included in wild landscapes, often nude, and surrounded by nature, emphasising their fertility and sexual potential. Women in landscapes were at once pure and fallen, passive and fertile. They were further linked to nature through the seasons, their bodies used to represent the passing of time and cycles of fertility. In other examples, women became nature, their bodies representing the landscape. The shapes of hills, the use of woods and flowers and the presence of water were included not only to symbolise the reproductive capacities of women but in the actual depiction of the female body.

Feminists have argued that such visual representations of landscape reflect male heterosexual desire and fantasy. They depict women as, like nature, sexually available and fertile. At the same time women are passive and submissive – again like nature, within the power of men. The ability of men to control women is linked to their ability to control nature and constitutes an important element of the wider gendering of the rural environment, as I discuss below.

While recognising the value of such readings of the landscape and the contribution they make to understanding the gendered nature of the rural environment, Catherine Nash (1996) argues that they need to be reassessed. She draws on images of the male body within the landscape and the work of female artists to question the notion that visual pleasure from landscape representation is necessarily simply masculine and heterosexual. Moreover, while that pleasure is always political, Nash suggests, it is not necessarily repressive. She writes:

> These images suggest that rather than simply assert the oppressive nature of images of feminised landscapes or of women's bodies as terrain, it is necessary to engage with them to disrupt their authority and exclusive pleasures and open up possibilities for difference, subversion, resistance and reappropriation of visual traditions and visual pleasure. (1996: 149)

Nash's paper introduces questions about visual representation and pleasure which are beyond the scope of this chapter. The importance of her arguments here is to warn of the dangers of generalisation in the recognition that visual representation of landscape can adopt gendered and sexual subject positions offering pleasures that are not limited to male heterosexual appraisal of women. Having said this, Nash herself acknowledges the importance of Western tradition in the history of landscape representation and recognises that the dominance of the male gaze cannot simply be erased or overturned through a reassessment of the politics of visual desire.

This section has concentrated largely on the gendered dimensions of *visual* representation. Similar points concerning the feminisation of the landscape, the male gaze and the association of women with nature could be made in respect to other ways of reading the rural landscape. Brace (1995) notes, for example, how a range of writings on the Cotswolds in England incorporates masculinist representations of landscape. She quotes, for example, Robert Henriques (1950) who writes of the Cotswold hills:

> [They] have another characteristic besides their invariable convexity; it is the gentleness and the two way nature of their slope. . . . [The hillside] is always rounded like the belly, breast or buttock of a living beast. To the eye it feels as soft, firm and resilient, as muscle and flesh. [The valleys] remain a sequence of folds in the hillside, so that to walk along a hill is to feel like a fly crawling from shoulder to shoulder across a woman's breasts. (Henriques, 1950: 31)

Again such writing displays not only the association between women and nature but also the values attached to, and reserved for, particular forms of femininity under the masculine gaze. Thus Henriques again writes:

> It is the valley itself which is the standard, not its component buildings, nor even its component villages. And when you visit this valley, you can savour it as you might a necklace of fine stones, you can either hold it at a distance and absorb the beauty and craftsmanship of its whole effect; or you can examine and admire each stone; or best of all you can hang it round the neck of a lovely woman to be filled with joy at the sight. That is the thing; to see the valley in its relation to the upland tillage, encircling with its glittering coils the windy hills and high barns. That is what matters *because however much you admire and envy a woman's jewellery, you cannot get any joy out of its beauty unless she herself has grace.* (1950: 72; my emphasis)

Examining literature on pre-1950s countryside recreation in England, Brace (1995) notes the association between order and chaos and gender in descriptions of the landscape. The Cotswolds as a successful and wealthy area of sheep production are portrayed as male but then, as agricultural decline sets in and the landscape becomes less controlled by farming practices, the 'language of description changes':

> By the twentieth century the image of the Cotswolds had been renegotiated by male observers who could not bring themselves to describe in masculine terms a landscape which appears to have reverted to feminised chaotic nature. (Brace, 1995: 12)

As the countryside is once again brought under control by agricultural development and recreational use the landscape is associated with masculinity. Significantly, the writing on the countryside at the time sees men, in the form of the walker/rambler, as the controllers of nature. Men are described as they walk in masterful fashion through the landscape and women are absent as men reassert a sense of order.

The next section of the chapter develops this idea of male control over nature in examining the relationship between masculinity and the rural environment. In so doing it provides a link to some of the points made in Chapter 7 regarding sexuality and the physical and social characteristics of the countryside.

Masculinity and the control of nature

In examining masculinity and the rural environment this section focuses on several different types of landscape including wilderness, productive countryside and forested landscapes. It also looks very briefly at rural landscapes as sites of tourism in showing how the relationship between masculinity and rurality is used in the consumption of different rural spaces. The section emphasises the complexity of men and women's relationship with the land and with the rural environment. It looks at ways in which the binaries employed in discussions about gender, nature and the environment can be destabilised.

Man's relationship with nature has been articulated perhaps most clearly in discussions of wilderness. Woodward (1998), for example, examines the ways in which British soldiers relate to the harsh environmental conditions of the remote countryside during their training, arguing that the ability to cope with such conditions is seen as an essential part of 'being a man'. She recognises that soldiers represent a particular form of hegemonic masculinity and that they are only one example of the relationship between gender and rurality; nevertheless the points she makes are valid and pick up on an albeit extreme form of the idea that an important element of masculinity revolves around endurance and stamina in the face of the natural environment. I return to this example in the discussion of the links between sexuality and rurality in Chapter 7 since the type of masculinity in question in debates here is, critically, heterosexual. Woodward (1998) quotes accounts written by soldiers of the hardships faced through exposure to remote, desolate and dangerous environments during training. She notes, in particular, the respect conferred by their colleagues on men who succeed in their training and are not beaten by nature. They have learnt about the countryside and how to survive in wilderness. Theirs is a 'rationalistic gaze' which sees the landscape as three-dimensional and understands how it fits together; such men are 'real' men and thus worthy soldiers.

A further interesting dimension to Woodward's argument is the representation of two different sorts of nature in the experience of soldiers. Reference is made, in army publicity, training documentation and the soldiers' own stories, to the (British) countryside that they are protecting and to the countryside in which their soldiering skills are tested. These, Woodward argues, are patently different types of rural environment. The first is the lowland, pastoral countryside ('bucolic and idyllic') in which people live and work, whereas the second is the wild, untamed and dangerous. Taming the elements of the latter to protect the former emphasises the power and control of masculinity over the rural environment. Wilderness is seen as a male space and, as Woodward (1998) points out, the soldiers' accounts of training in such environments include not only descriptions of the danger and physical hardship but also celebrations of the comradeship and sense of adventure that these exercises evoke. The military training exercises in which soldiers are required to survive for several days in the wilderness are likened at times to boys' adventures.

The wilderness as a place for boys' adventures is a theme addressed by Richard Phillips (1995) in his accounts of Victorian masculinity. Phillips shows how Victorian writers situated boys' adventure stories in wild, unpopulated, primitive landscapes in which their manhood is tested and defined. The boys and young men of these adventure stories are required to perform feats of physical bravery and prowess within the landscape – walking long distances through snow, scaling mountains and swimming across rivers – to demonstrate and confirm their masculinity. As Phillips points out, a highly simplistic universal masculinity is 'naturalised' in these texts; and this dominant form of masculinity is made more plausible by the 'simplified, caricatured ruggedness of the setting' (Phillips, 1995: 604).

As with Woodward's military examples, the rural landscapes of Phillips' Victorian adventure stories are resolutely male spaces. The wilderness is populated almost exclusively by men and where women do appear, they are largely silent and passive. Women are objects, as Phillips observes, to be:

> variously killed, fought over and saved. They are vehicles through which men define themselves. For the most part, though, women play little or no part in the adventure which is an all-male affair. (1995: 601)

Men's masculinity is defined, therefore, in these wilderness settings, in relation to other men and in the absence of women. Having said this, the presence of nature constitutes, as Phillips notes, a form of unacknowledged feminine other: 'the metaphorical femininity of the landscape presents an other against which the hero can define his masculine self' (1995: 601). Both Woodward and Phillips cite examples of the portrayal of nature as female in their discussions of the relationship between masculinity and landscape, emphasising the pervasiveness of such feminine archetypes in constructions of nature. The wilderness incites constructions of nature in which women are not warm, helpless and gentle but cold, forbidding and mysterious. Nature can be fickle and treacherous and even when men 'penetrate' nature they do not get too close for fear she will 'turn without warning' (Phillips, 1995).

A further example of the relationship between wilderness landscapes and masculinity is provided by Cloke and Perkins (1998) in their study of adventure tourism in New Zealand. These authors look at the increasing popularity of a range of adventure activities in the tourist experience of New Zealand, showing how such activities (including bungy jumping, white-water rafting and jet boating) are predicated on a particular construction of the natural environment. They are about taming, and overcoming the challenge of nature while at the same time being humbled by the beauty and power of the environment. The adventure tourism experience in New Zealand then means:

> exploration of uncharted territory, experiencing the danger and adrenaline rush of past explorers; travelling the untravelable; seeing the unseeable; generally pitting adventurousness, personal bravery and technological expertise against natural barriers and winning. (Cloke and Perkins, 1998: 204)

The portrayal of the landscape conforms to the masculine gaze in which nature is to be conquered and penetrated but also revered and respected. The adventure tourists, like the soldiers of Woodward's paper or the explorers discussed by Phillips, are a little distanced from nature, mindful of her fickle and dangerous potential. While Cloke and Perkins point out that women as well as men participate in new forms of adventure tourism, they recognise that the leaders/instructors of activities tend to be male. Moreover, whether consumed by men or women, the images of adventure tourism are undoubtedly macho. They portray fit and healthy bodies and the activities themselves largely exclude all but the able-bodied. This is an elite type of tourism offering a particular form of interaction with the natural environment and embodying stereotypical gender images, although less exclusionary of women.

The naturalisation of hegemonic gender relations within wilderness landscapes is explored in some depth by Alastair Bonnett (1996) in his study of the 'mythopoetic' men's movement.[2] According to Bonnett, the mythopoetic men's movement draws on an intellectual heritage that he

---

2   The mythopoetic men's movement first emerged in North America in the 1980s. It celebrates a form of masculinity which reaffirms male power and is essentially hostile towards women and, in particular, feminism. The movement has been heavily influenced by Robert Bly's book, *Iron John* (1990). For a fuller explanation see Bonnett (1996).

summarises as 'wilderness philosophy', in which nature or wilderness affords men an escape from 'the claustrophobic, alienated and artificial world of bourgeois/feminine society' (Bonnett, 1996: 282). Bonnett describes the weekend 'retreat' where members of the movement camp out in the wilderness performing ritual ceremonies which draw on the mythical terrain of nature to naturalise male power. The natural landscape is seen as the appropriate place for natural gender relations, but this landscape is:

> not simply a place where men rule. It is a less crude but, perhaps, more insidious destination. For the mythopoetic movement is focused upon a space, a time, where men and women's essential character is 'recognised', where the eternal struggle between feminine and masculine energies is 'understood'. (Bonnett, 1996: 285)

The mythopoetic men's movement may be a relatively minor social movement and yet it displays an interesting and powerful mobilisation of ideas of nature and wilderness in explicit defence of highly traditional forms of masculinity. Moreover, these are elements of masculine identity that are perceived as under threat from the feminist movement and by changing gender relations. The retreat to nature allows values of timelessness and stability to be employed in support of threatened masculinity.

The relationship between masculinity and control of nature is arguably not as apparent in other types of (non-wilderness) rural landscape. Studies of farming (see Brandth, 1995; Bryant, 1999; Saugeres, 1998) and forestry (Brandth and Haugen, 2000), however, have suggested that powerful images of masculinity are incorporated in attitudes towards the land and in the valorisation of different parts of the production process. Lise Saugeres (forthcoming) suggests that for the farmer land has a symbolic as well as an economic value and that the land is 'the means through which men create their masculine identities'. But she then goes on to argue that men have, in fact, a highly contradictory relationship with the land in which they are both apart from and, at the same time, connected to nature. This contradiction is interesting as it is also picked up by Jones on his work on gender, nature and childhood, as is discussed below.

Saugeres uses research conducted amongst French farmers (see Saugeres, 1998) to illustrate her arguments concerning masculine identity and nature. She carried out detailed interviews with 36 farm families in Southern France to build up life histories of women and men in relation to their changing involvement with the farm business and with the land. Saugeres makes several important points concerning the links between gender and responses to the land. She recognises, for example, the presence of a masculine gaze in which nature is seen as female and men's relationship to the land one of dominance and control. Comments from male farmers describing the pleasure they obtain from working the land are used to illustrate the power associated with routine agricultural tasks such as ploughing and sowing. The land, it is argued, is both Mother and object of desire (as is also discussed by Kolodny, 1975); the farmer is not only connected to but also separated from the land in a way which echoes men's relationships with their mothers and their lovers.

The power of the male farmer over the land can also be threatening and destructive. Here again, Saugeres argues, man's relationship with the land parallels patriarchal ideologies between the husband and the wife. Man's mastery over the land must be treated with caution – if he does not take care of his land and animals they may rebel against him. This dependency requires that the farmer looks after his land (and wife) or it will not provide for him. Thus dominance and control also implies distance and respect. In addition, Saugeres discovers, the relationship between farmers and their land is not seen to be something that is automatic. The ability to understand and control the land is believed, amongst farmers, to be something that is usually passed down through generations. It is not necessarily something that can be learned but is something that is in their blood. As Saugeres thus sums up:

> The dominant ideology of farming is articulated around masculinist ideas of the land and nature and the idealisation of the past as a natural space. (forthcoming)

The relationship that Saugeres observes between women and the land in her study of French farming families is very different. She notes how women talk of the land as men's space and of working the fields as men's work. While men construct their farming identities in relation to the land and nature, women do so in relation to their husbands. Women, then, are peripheral to the running of the farm – even in cases where women farmed alone, they were perceived to be doing so in the absence of a man – and do not have the same contact with, or feeling for, the land. Even though women are seen as 'closer to nature', it is the control of nature and the ability to understand what is happening on the land that define men's identities as 'natural' farmers.

In examining the relationship between gender identity, nature and the land in the context of farming, Saugeres identifies what she sees as contradictions: men are seen (and see themselves) as both masters of nature and at the same time deeply connected to the land. She suggests that this contradiction is articulated in the opposition between traditional peasant farming and modern agriculture. This assertion highlights the role of technology in the construction of gendered farming identities – an area on which other authors have focused (see, for example, Brandth, 1995; Brandth and Haugen, 2000).

The belief that the peasant, the farmer born into a farming family, is a 'better' farmer stems partly from the idea that such men are in some way 'closer to the land' than men that have come into farming later in life. While the peasant farmer might have access to particular, traditional, farming skills and knowledge, however, there is recognition of change in the nature of agricultural production. The relationship between farming, male identity and land is also overlain by that of masculinity and technology and by the contradictory notion of control. Use of technology allows men to better control nature and the rural environment but at the same time may distance them from the ancient skills of the peasant farmer. Brandth (1995), however, argues that changing masculinities together with shifting ideas about the appropriate and effective use of machinery in agriculture has meant that technological knowledge has become an important, and highly valorised, aspect of masculine identities in farming.

Brandth (1995) looks in some detail at the relationship between masculinity and technology in agriculture in her study of visual representations of tractors and farming in Norwegian advertising. She argues that such representations provide a valuable contribution to social analysis that goes beyond the use of written text alone. Brandth draws attention to the emphasis placed on the size and power of tractors:

> The tractor is pictured as large and strong. Big, heavy, greasy, dirty and noisy machines are symbols of a type of masculinity which is often communicated through the male body, its muscles and strength, and what is big, hard and powerful. (1995: 126)

The ads contain no women and depict farming as a lonely occupation. This accentuates the relationship between man and machine and between man and nature. The value of the tractor is in its power to control nature. To quote Brandth again:

> Contemporary, industrialized farming practices are heavily dependent on technologies like the tractor and other machinery. The control of the farmer and his machine over nature, as . . . symbolised in the adverts, is

very much in line with masculinity expressed in the scientific tradition that sees progress as a matter of domination and control over nature. (1995: 128)

The introduction of ever more sophisticated technologies in tractor design reflects, so Brandth argues, changing masculinity within farming. The emphasis is directed away from heavy manual work towards a more 'white collar' image. The use of science to control nature has produced a different sort of masculine identity within farming but although the shift away from pure physical work has taken place, this has not been reflected in the gendered nature of farm labour. Thus women remain largely absent, and farming and technology remain important areas of male work.

Similar sorts of changes in masculine identity are also evident in the representation of a related rural industry – forestry. Brandth and Haugen (2000) show how changes in the production process, including the introduction of technology, are reflected in the masculinity of forestry workers. Forestry in the 1970s and 1980s is depicted as an extremely macho activity. Work is physically demanding, dirty and dangerous; moreover, it is likely to take place in harsh conditions. Again, the ability to control nature through strength and fitness is a defining feature of rural masculinity amongst those working in the industry. As with agriculture the ability to use heavy machinery is also a significant feature of masculinity as portrayed in the case of the forest worker. Over time, however, the increasing use of computer technology to control such machinery has tended to challenge the 'tough man' image of the forestry worker, who now requires new skills to control nature.

The ability to control nature in the forestry industry has shifted subtly and is seen now to embrace a 'new' form of hegemonic masculinity which Brandth and Haugen (2000) dub the 'organisational man'. This is the new manager whose success is defined not by prowess in the forests but by economic results and competence in the boardroom.

The power of the organisational man is based on control over economic resources as he leads and manages other men's interests, displaying masculinity by means of the 'power look' of business suits, conference tables and rostrums. (Brandth and Haugen, 2000: 354)

Brandth and Haugen (2000) argue that both 'versions' of masculinity within the forestry industry depict men doing battle – the traditional forest worker with the natural elements and the new manager with the stock market and world of business. In a sense, however, both are involved in the control of the environment and the transformation of the rural landscape. Shifts in the construction of masculinity encompass the fact that such control can incorporate administrative ability and not simply physical strength.

Despite the changes in the masculinity of forestry workers, like agriculture, forestry remains a strongly gendered occupation: women are mainly absent from representations of forestry, they appear only infrequently in either the woods themselves or in the boardrooms (see Brandth and Haugen, 1998, 2000). Attempts to encourage women into the Norwegian forestry industry in the 1980s did have a positive effect on the numbers of women forestry workers but did little to challenge hegemonic male discourses, and the portrayal of forestry work remained highly macho. It was clear that women were present as 'tokens in a man's world' (Brandth and Haugen, 1998: 439).

There are very few studies of the relationship between gender identity, nature and rurality as it is played out within the village itself. Bell's study of Childerley, however, makes reference to the seemingly different relationships that men and women have with nature and how these are manifest in practices and customs of the rural community. He notes the gendered patterns of interaction with nature, drawing attention to men's greater involvement in outdoors work – especially that involving dirt and physical strength. He speculates as follows:

> my guess is that this kind of dirt of outdoor work and leisure (and not of poverty) is a complex emblem for them (the male Childerleyans). It is, I suspect a sign of mental toughness, for it shows an ability to overcome fear of the outdoor threat to physical comfort. It is probably also a sign of physical toughness, for it presents evidence that a dirty man has tried his bodily prowess against the outdoors. (Bell, 1994: 214)

Another sign of this male interaction with nature comes, he also suggests, in the form of killing outdoor animals – a task which is mostly undertaken by men. Bell goes on to suggest that this kind of relationship does not exist between women and nature because of women's 'more domestic experience of the self'.

So far in this section I have discussed examples of the links between masculinity and the rural landscape as represented through ideas surrounding the control of nature. While these examples have provided clear and interesting evidence of the ways in which rural masculinity is tied in with mastery of nature, many of the studies referred to stress inconsistencies in the relationship between masculine identity and the rural landscape. Both Saugeres (1998) and Bonnett (1996), for example, note the tension in the representation and performance of masculinity between the ideas of men being in control of nature, and thus distanced or apart from the landscape, and being close to or part of nature. Although changing masculinities do suggest – as demonstrated by Brandth (1995) and Brandth and Haugen (2000) – shifts in the relationship between gender identity and nature, such changes do not fully explain the broader inconsistencies. Moreover, there are also tensions between feminised nature as chaotic, helpless and controlled by men and as powerful, unpredictable and dangerous.

The tensions in the relationship between nature and constructions of masculinity and femininity within the rural environment are articulated by Owain Jones (1999). His research on 'country childhoods' has examined the different responses of boys and girls to the rural landscape and the differing expectations of parents in relation to the performance of masculinity and femininity in the countryside. Jones argues that the gender of childhood in close harmony with the natural world is 'necessarily constructed as male' (1999: 118). He looks in depth at the relationship between the countryside and children's lives in Britain, noting the strong and persistent role of childhood in constructions of rurality and, in particular, in the rural idyll. Literary and lay discourses are employed to demonstrate the association of 'rural' images of freedom, purity and innocence with childhood. Jones draws particular attention to representations of the countryside as a place of healing for children and of nature as an anecdote to the disease and impurity associated with the city (see also Jones, 1997).

The gendered stories described by Jones (1999) portray the activities and spaces of rural childhoods as mainly male. The play is more adventurous, scruffy and physical in a way that is more often associated with, and tolerated in, boys. Jones cites a number of examples of childhood literature in which the rural adventures that take place are almost exclusively male – girls are treated as an intrusion, their presence serving only to threaten or spoil the boys' adventures. Where girls are successfully incorporated into the play or space of countryside stories it is as tomboys – quasi, honorary boys. This difference in the relationship of boys and girls to the rural landscape is due, Jones suggests, to the construction of male children as 'natural' and able to be at one with nature.

This idea is further supported by fables of 'wolf children'. In the numerous recorded accounts of children being raised by wild creatures and/or living 'wild' in the most literal sense, it was mostly male children who were found . . . But in the accounts which became more widely known . . . famously fictionalised as in Tarzan and, of course, Kipling's tales of Mowgli, it is male children who can not only be close to nature but actively become part of nature. (Jones, 1999: 128)

As noted, this naturalness of masculine identity in rural childhoods presents some problems for the romantic constructions of nature as 'female'. The 'problematic relationship' between gender, childhood and nature is further complicated, as Jones also points out, by the attitudes surrounding the development of female sexuality. While it could be argued that girls' sexual maturity brings them 'closer to nature', in the case of countryside childhoods it appears to distance them from 'natural' spaces and pastimes. Even the tomboy identity and the access to nature as a quasi male are no longer available to girls as they mature. The ways in which gendered rural childhoods are constructed suggests, Jones argues (1999: 133), that while male children are seen as part of nature and 'grow into cultural beings', female children travel in the opposite direction, starting as entities of culture who then 'transform into natural beings'. Why this happens, he speculates, is because the 'ground of childhood . . . is too valuable a ground for patriarchy to concede' (1999: 132).

Women, nature and the rural environment

So far this chapter has examined the construction of masculine and feminine identities within the rural landscape and shown how these relate to ideas of nature. Particular attention has been directed towards debates surrounding the control of nature and the links between hegemonic masculinities and mastery of the rural environment. I now want to turn to look at the implications of the relationship between gender identity and nature for women's use of the rural landscape and for their broader connection with the environment through political activism. It is not the intention here to examine in detail the background of the environmental movement nor to debate the philosophical roots of eco-feminism. Rather, it is simply to show how some commentators have employed women's involvement in the environmental movement in further uncovering, and illustrating, the relationship between femininity, gender identity and nature.

Many of the studies discussed in connection with masculinity and the control of nature remarked on the absence of women from the rural landscape – especially from wilderness areas. Despite the 'closeness to nature' of constructions of women's gender identities, they are not always seen as 'in place' in the rural environment and there are mixed responses, from both women and men, to women's use of and presence within the countryside. While women are increasingly involved in recreational and sporting activities that take them into 'wilderness' environments (see Cloke and Perkins, 1998), their presence there may be seen as unsettling and inappropriate, for reasons discussed below. Alternatively, remoter rural areas can be seen to provide the space for women to live apart from men and to live in communities that put them in touch with nature and, consequently, with their own femininities.

Bialeschki and Hicks (1998) have explored the issue of safety amongst women taking part in outdoor recreation, noting the extent to which women's fears curtail their involvement in activities in remote environments. These authors carried out in-depth interviews with 21 women, all of whom regularly engaged in outdoor recreation. All the women interviewed claimed to feel afraid at some time during their activities although most said that they were determined not to let this fear curtail their participation. Interestingly, when questioned further about their fear, the women stressed that while they may be more afraid in remoter areas it was not the environments themselves that they feared but the people (men) within them. Talking about hiking, for example, one woman was reported to have said:

> My (fears) have more to do with things like people attacking me than like forces of nature (quoted in Bialeschki and Hicks, 1998: 4).

Another woman backpacker explained:

> We ended up in a shelter and the shelter had 8 ATV (All Terrain Vehicle) routes up to it and it had roads where you could see someone in ATVs had come up there. There were signs of tequila bottles and beer cans and stuff that actually made me feel very insecure . . . The next night we were in an area that was far away from accessibility and I didn't feel awkward or insecure at all (quoted in Bialeschki and Hicks, 1998: 7).

While this is obviously a small study and makes no claims regarding representativeness, it does highlight important issues concerning not only the fears felt by women in going into remote rural environments but also the source of their unease. In this instance it was not nature itself that bothered women but what nature might be concealing. As has been found in research on women's fear of attack in urban environments (see

Valentine, 1992) the women in Bialeschki and Hicks' study talked of the precautionary measures they adopted to reduce the likelihood of attack. The comments by some that they were foolish to exercise in certain places and at particular times since it increased their vulnerability reflected the view that there are some environments that are 'out of bounds' to women. While women could be skilled and practised in coping with the dangers of the natural environment, they were still at risk from a human element. As the authors sum up:

> the potential for male violence in certain spaces has a profound impact on women's activities to the point where women 'take for granted' that they may be limited in their choice of routes, destinations and companions. For the women in this study, they would at times develop options to extend their choices but generally seemed to recognise a finite boundary where their anxiety and discomfort would exceed any potential benefits from the activity. (Bialeschki and Hicks, 1998: 11)

Concerns about the appropriateness of women in remote rural settings are also apparent in reports of women being attacked. In recent cases in Britain of women and girls attacked while out walking in rural England the press have suggested that there has been a dramatic shift from the countryside as a safe space to one of danger to women on their own and that women should appreciate this shift in planning their recreation. The sense in which lone women walking in urban areas have been seen as 'asking for trouble' has started to be applied to reports of violence in the countryside. Women clearly belong in the spaces of the rural community but not in the spaces beyond unless accompanied by a man or as part of a larger group.

Tim Cresswell (1996) develops this theme of women as 'out of place' in the natural environment in his discussion of the 1980s women's 'peace camp' at Greenham Common. He notes that while women's supposed 'closeness' to nature was employed in affirmation of their unique role in protesting about the potential damage caused by nuclear weapons to the land, the environment and to society (especially its ability to reproduce itself), the ways in which women were living in the camp – in particular the lack of facilities for washing and cooking – were seen to conflict with their femininity. Women's presence was seen as inappropriate in an environment which in all its naturalness was perceived as squalid and unhealthy, despite the centrality of the protection of nature to their protest.

Having put forward some examples of situations in which the relationship between women's gender identity and nature serves to render them 'out of place' in the rural landscape, I now want to suggest some alternative cases in which the association between femininity and nature is used to enhance women's experience of rurality. The first of these concerns the particular formation of separatist women's communities in the United States while the second relates to a broader issue, namely the involvement of women in environmental politics. These very different cases are used to show another, more positive reading of the links between women and the rural landscape.

The creation of separatist, lesbian communities in the USA is discussed by Valentine (1997) from her readings of Joyce Cheney (1985). The communities were established in the 1970s with the aim of creating a new society beyond the influence of men. A few were formed in urban areas but most located in the countryside where it was easier for women to be self-sufficient. The specifically lesbian dimension of these communities is discussed in Chapter 7; the (not unrelated) relevance here is in the perception of the countryside as 'women's space'. Essentialist notions about women's relationship with nature and the land stemming from their reproductive capacities and the cycles of their bodies were mobilised to claim the rural environment as the most appropriate place for the expression of feminine identities. Women also idealised the countryside in a political way as Valentine notes, imagining it as 'simple, peaceful, safe space untainted by patriarchy' (1997: 111).

Nature, in this context, is seen to *protect* women. As such it was celebrated in the communities in the ceremonies and practices observed. There was an emphasis in the communities on women growing their own food and relearning other skills closely associated with nature such as herbal medicine and fire-making (see Valentine, 1997). In some ways the construction of the rural environment as female space seems to contradict the earlier examples of women feeling out of place in remote countryside. However, it is the presence of men, in both cases, that disrupts the relationship between femininity and nature; in the example of women's outdoor recreation, their fears relate to attack by men while in the separatist communities, strength in nature is drawn from the absence of men. When, as Valentine (1997) describes, the communities of the 'lesbian lands' began to experience tensions and finally break down, it was the response to outside influences and the invasion of, in particular, urban ideas and practices that were to blame, not the relationship between women and nature.

The relationship between women and nature has also been used as a 'motif for mobilising women's environmental activism' (Lipiens, 1998: 1183) (see also Brandth, 1994; Sachs, 1994). Some more radical

eco-feminists have, for example, drawn on the women–nature relationship, and the idea that women's reproductive and nurturing activities
together with their spirituality place them closer to nature, to argue for
women's involvement in environmental politics. They claim (in accordance
with the binaries operating in western thought discussed in Chapter 2)
that a link exists between male domination of women and of nature and
that men's treatment of the environment is the basis of their oppression
as women. Thus as Sachs claims, eco-feminists argue that:

> As an outcome of their subordination, women have a stake in ending and
> challenging the domination of both nature and women and recognise
> that solutions to ecological problems must be tied to social and gender
> transformations. (1994: 118)

This perspective has been criticised for essentialising the relationship
between women and nature in a way that is seen as potentially regressive
in terms of the broader politics of women's oppression (see Biehl, 1991;
Lipiens, 1998; Monk, 1992). Other eco-feminist positions (known vari-
ously as social eco-feminism or feminist environmentalism) argue that the
close association between women and nature lies not in any 'fundamental
feminist traits' but in the 'constructions of gender and the social positions
in which women are placed' (Lipiens, 1998: 1183). It is argued, therefore, that women's valuable contribution to environmental politics and
activism stems from their knowledge of the environment and from their
experience as carers and as managers of (often) scarce resources.

Women's involvement in particular environmental campaigns has been
documented by other rural social scientists. Sachs (1994), for example,
has looked at the position and experience of women in the sustainable
agricultural movement in the USA. She has also examined the role of
women in campaigns surrounding toxic waste and pollution. Lipiens
(1998) has investigated the environmental concerns and activism of
women in relation to agriculture in Australia. In Britain, Buckingham-
Hatfield and Percy (1998) have explored women's involvement in local
Agenda 21 environmental projects, noting in particular the role played
by the gendered governmental and organisational structures in women's
participation.

It is not the intention to go into the detail of this research here. What
the various studies have demonstrated, however, is that women are increasingly willing to participate in environmental debates and to use
their skills and knowledge to speak out on environmental issues. Lipiens
(1998) has focused on the different subjectivities adopted by women in

terms of their involvement in environmental activism in Australian agriculture. She claims that women identify as farm workers, business women, partners, mothers and carers and that:

> these positions enabled women to speak on environmental issues based on particular constructions of nature, rural space and womanhood. (Lipiens, 1998: 1192)

She also warns, however, that the adoption of the position of mothers/ carers in relation to environmental debates could result in the exploitation of women's labour in making them accountable 'not only for the unpaid and underrecognised social reproduction of families and communities but also for the nation's socially nurtured and well-ordered rural environment' (Lipiens, 1998: 1192).

On a rather different but related issue, Brandth and Bolsø (1994) have examined the gendering of attitudes towards biotechnology and nature in the production of food. They again locate their discussion within the theoretical framework of eco-feminism but include within this a feminist critique of masculinist science and technology. They question whether men and women hold different moral attitudes toward biotechnology and if so, under what circumstances do such differences exist? Brandth and Bolsø conclude from their research that while women and men do hold different views on the ethical implications of biotechnology as consumers (with women being more concerned than men about the use of genetic engineering in particular), these differences disappeared in the case of farmers and scientists. As women's and men's contexts become more similar, they argue, their views tend to become more alike. Brandth and Bolsø (1994) could only speculate, on the basis of their research, about the implications of these findings for the relationship between gender and the environment but clearly it does potentially disrupt essentialist ideas surrounding the association between femininity and nature.

Conclusion

The relationship between gender and nature underpins many of the values, assumptions and attitudes that surround women and men's experiences of rurality. In this chapter I have tried to show how our views of the rural landscape and of who or what is appropriate within that landscape stem from, in a very fundamental way, the association between gender and nature. The ability to control nature in both wilderness and farmed rural landscapes resonates with traditional hegemonic masculinity and with men's power over women. And while men 'rise above' nature, women are seen as much more firmly embedded within, controlled by, and part of nature.

The relationship between gender and nature has been illustrated here through various aspects of men and women's lives from recreation to agriculture. The chapter has drawn heavily on published work in an effort to pull together the key ideas and open up important areas for debate – areas that will inform the following chapters. The examples discussed have shown some of the practical manifestations of the centrality of ideas of control to the links between masculinity and nature. They have also demonstrated something of the complexity of the relationship between gender and the rural environment. The key point perhaps is that nature and the rural landscape incorporate and reflect gendered power relations; power relations, moreover, that are not confined to the way we experience nature and the rural landscape, but impinge on other social and economic characteristics of the countryside through their role in constructions of masculinity, femininity and rurality.

4 | Gender and the rural community

Introduction

Chapters 1 and 2 have both made reference to the importance of ideas of community to the construction and articulation of rural gender identities. The highly traditional nature of gender roles and gender relations has been noted as a key characteristic of rural society and one which both draws on and contributes to dominant constructions of community in the countryside and to the ways in which such constructions are played out in the daily lives of those living there. Understanding the rural community as not only a 'container' for the operation of gender roles but also as a socio-cultural factor in the evolution of gender relations is essential to this study of gender and rural geography. Encapsulated in the practical workings of the rural community and in the meanings which circulate around it is a powerful set of values and assumptions about how women and men relate to each other and make sense of their lives.

The community clearly constitutes an important site for the organisation of gender roles and as such a valuable focus in terms of the documentation of the detail of men and women's rural lives. In this chapter I look at the variation between women and men's activities within the rural community; their differing levels and forms of involvement in the community as a social institution and the importance they attach to this involvement. In doing so, the chapter draws on various pieces of research, undertaken by myself and by others, in an attempt to establish how relevant are gender divisions to participation in the community – in its activities and in its running – and to the assumptions surrounding involvement in and responsibilities towards the rural community.

Discussion will move from the organisation of gender roles within the rural community to the broader theoretical context and to the relationship between gender identity, gender relations and the cultural construction of the community. Here I examine the meanings and values attributed to the rural community and how these reinforce particular gender identities and encourage traditional gender relations. Particular attention is given here to the relationship between the domestic sphere and the rural community and to the extension of women's domestic responsibilities as an essential element of both the real and the imagined community. Integral to such an analysis is the question of power within the community, and this chapter will introduce a number of ideas concerning the gendered nature of community power which will be taken up again more directly in Chapter 6.

Throughout the various sections of the chapter I will consider ways in which the notion of the rural community has been contested. The strength with which the dominant constructions of community are held both inside and outside rural areas means that examples of 'alternative' constructions that challenge the accepted views are few and far between. Thus I look here not so much at communities that have been organised in a different or alternative way but rather at groups within the rural community for whom the dominant constructions are inappropriate or alienating. Such groups may not have the power or influence to disrupt the conventional interpretations and their material and ideological implications for gender identity, but may be able to launch specific (possibly single-issue) challenges to the dominant constructions.

The community in rural geography

Rural researchers from a range of disciplines, including sociology, history, anthropology and geography, have long been fascinated by the rural community. Indeed as Lipiens (2000a) and Wright (1992) have observed, research on various aspects of the rural community has been substantial and significant enough for the identification of a specific genre of 'community studies'. Research on the rural community began in earnest just after the Second World War. Through the 1950s and 1960s there was a steady supply of such rural community studies – generally focusing on a single village and providing a detailed description of life for all members of the community from the major landowners to the agricultural labourers (see, for example, Frankenberg, 1966; Littlejohn, 1963; Williams, 1956). While these studies provided rich and valuable data on the nature of rural life, they attracted criticism for their highly static perspective in which communities were viewed as 'relatively discrete and stable phenomena with observable characteristics (structures) and demonstrable purpose' (Lipiens, 2000a: 24). The social relations within them were seen as fixed and unchanging, and communities existed in some kind of functional equilibrium.

These early rural community studies were problematic, not only in terms of their functionalist perspective but for their treatment of the relationship between social characteristics and place. Traditional rural community studies made the fundamental mistake (according to Gans, 1962; and Pahl, 1965) of appearing to imbue the rural community itself with causal powers, reifying the concept of community as an 'active social entity' (Day and Murdoch, 1993: 8). As geographers especially became sceptical about the role and integrity of space, the value of rural community studies began to be questioned. As Day and Murdoch argue:

> For many commentators, the rejection of the usefulness of 'community' was part of a much broader onslaught on all forms of explanation which attached causal significance to space, and to spatial relations. (1993: 84)

The theoretical and conceptual doubts surrounding the rural community led to a decline in the rural community genre during the late 1970s and 1980s. This, however, left rural researchers with something of a vacuum and rural studies, arguably, lost a sensitivity to the richness of people's lives and to the interconnectivity of aspects of social and economic relations in particular places. During the 1980s some attempt was made by geographers to recapture something of the role and value of place in the development of what became known as 'locality studies'. While such studies did seek to uncover local detail and place-specific characteristics, they remained consistently wary of the notion of 'community'.

The emphasis they placed on global processes of restructuring further detracted from the value of a community-based approach and fashion for in-depth rural community studies continued to recede. Reference to the rural community as either a geographical or a social entity became somewhat 'minimalist' (Lipiens, 2000a), particularly in the face of political economic approaches and the search for global patterns and processes of socio-economic change. Within this 'minimalist' phase some rural writings 'hardly refer to "community" at all [while] others briefly employ the term to denote a scale of inquiry or a loosely specified sense of social collectivity' (Lipiens, 2000a: 25). Similarly, community has been used by rural researchers as a methodological descriptor – essentially to provide a container for conducting and reporting research.

Wright (1992) has identified another approach to the study of the rural community that developed alongside this 'minimalist' perspective: the 'symbolic construction approach' in which emphasis is placed on the meanings and symbolism attached to the use of the term 'community'. Such an approach moves away from the idea of the rural community as a fixed and stable entity which occupies an identifiable geographical location. In particular it questions the differential experience of belonging and of social cohesion as key features of the village community (Lipiens, 2000a). While proving valuable in drawing attention to an important way of conceptualising the rural community, the symbolic construction approach is not without its shortcomings. In particular, used in isolation, it provides insufficient scope for the identification of material practices involved in the construction and operation of community. There is, moreover, a downplaying of the role of power relations – especially in terms of the differential ability of individuals to influence and control specific symbols or meanings of community. Such an approach does, however, begin to raise issues of individual experience of community that have been picked up again more recently.

Cultural and postmodern approaches in geography and the social sciences have reinvigorated rural community studies, providing a context for a re-engagement with ideas surrounding the spaces, meanings and practices of the rural community. As Lipiens (2000b) notes, new directions in the study of the rural community have begun to emerge in the past five or six years in which the socially constructed nature of rurality is acknowledged. As a result the rural community has been conceptualised as a more complex, fluid and contested entity. Studies have been directed towards the identification of difference in the experience of community rather than sameness and to exclusion rather than inclusion (see Cloke and Little, 1997; Milbourne, 1997). Such work has drawn on the arguments of Young (1990) in re-examining and problemising the 'ideal' of community and in the recognition of the negative experiences of those whose identity may serve to exclude them from various practices and structures of community.

As well as allowing a focus on different experiences of the rural community, postmodern approaches in rural studies have also encouraged work on the cultural construction of the rural community and on the role that such constructions play in our understanding of rurality itself. The study of different representations and discourses of rurality has identified the key place of ideas and meanings of community in popular imaginings of the countryside. It has also emphasised the power relations encapsulated in the articulation and negotiation of dominant notions of community within rural society (see Crouch, 1992; Jones, 1995). These recent developments in rural studies have provided considerable scope not only for a re-engagement with the idea of community but also for continued investigation into the different ways community is experienced. In the context of this book, current interest in community both invites and supports the study of gender difference. Emerging from recent work is a clear need to investigate not only the ways in which different experiences of the rural community are gendered but also how gender relations are embedded in the constructions and meanings of community and within the power relations through which these are sustained.

Before turning to look directly at gender and the rural community it is worth noting the four different dimensions of community that Lipiens (2000b) proposes as a framework for reading and understanding the social formation (and maintenance) of the rural community. She argues, first, for a focus on the people within a community to encompass an appreciation of the 'collective interaction and enactment of "community"' and the varying positions occupied by people both inside and outside the rural community. Secondly, attention needs to be given to the 'shared meanings' surrounding community and to the articulation of these in narratives and texts. Thirdly, the practices through which people come to construct and enact their understanding of community must be considered. Fourthly, Lipiens argues that the spaces and structures through which these practices are played out should be included in attempts to understand the rural community. She shows how using these dimensions as a framework for reading the rural community in its material and immaterial forms can lead to an understanding of the rich diversity of rural social change and power relations.

Rural community studies and the recognition of gender difference

Past studies of the rural community, while arguably theoretically and conceptually flawed, nevertheless provided an important vehicle for the identification of gender differences. The focus on lives of individual agricultural workers and on the collective experiences of the rural household contained in the early 'community studies' referenced above, provided an insight into rural gender roles. Through such studies an understanding of the gender division of labour on and off the farm began to develop and while, as noted in Chapter 2, studies were mainly of a descriptive nature, they did draw attention to differences between men and women's lives in the rural community. In particular the key (but largely unrecognised) roles of women as agricultural labourers and as domestic workers were highlighted as central to the existence of farm and village life.

Even though 'community studies' declined as a genre, the community retained an important role in the development of feminist rural geography and in the understanding of gender roles and relations. There is insufficient space here to provide a comprehensive examination of the various community-based studies of gender difference. It is useful, however, to draw attention to some of the main points that have been made by such studies and to provide some illustrations from the various case studies on gender and the rural community that have been undertaken.

Sue Stebbing (1984) provided early evidence of different gender roles in the rural community, and in particular the activities of women, in her research on parishes in Kent in South-east England. In her study Stebbing drew attention to the relationship between service provision within the rural community and women's gender roles. She argued that the relative remoteness and the poor level of resourcing of rural communities reinforced the gender division of labour and emphasised women's roles. The practical difficulties involved in carrying out everyday domestic tasks served to complicate women's roles and to confine them to the home and community. Stebbing also argued that the lack of formal provision of some basic health and community facilities (especially care for the elderly, childcare services and social clubs) was more readily tolerated in rural communities where gender roles were more traditional and the expectation was that such work was part of women's daily responsibilities.

> Where there are no social services provided by the State the burden of unpaid care for the elderly, the sick and the pre-school child usually falls upon women. Women are seen as the 'caring' sex for whom it is little or no hardship to take on additional nursing or nurturing responsibilities at home since that is their primary role in any case. A local psychiatric social worker reported that a significant difference between urban and rural referrals to her psychiatric hospital was that many rural referrals were senile women admitted in order to relieve relatives (often daughters) who had no access to day care facilities. (Stebbing, 1984: 10)

As Stebbing makes clear, the caring role 'naturally' performed by women extends almost automatically from the home to the community. A whole range of community tasks is pigeon-holed as an extension of women's domestic role and these have become gendered in terms of both involvement and responsibility. This is a theme taken up more recently by other authors. Hughes (1997a, 1997b), for example, examines the lives of women in a rural community in mid-Wales (which she names 'Ditton' in the study) and notes their pivotal role in the organisation and operation of the community. Women, she argues, were much more prominent than men in the whole range of community activities, from essential caring duties to the running of social events such as the village fête and annual bonfire. Hughes again stresses the expectation of women to 'do their bit' in the rural community. As she sums up:

> The gender division of labour in the organisation of village events is clear. Women were the backbone of the village community and, moreover, they were expected to carry out community work. Community participation was perceived to be their duty, their natural role. (Hughes, 1997a: 182)

Evidence has been presented from studies in a range of different localities, both in the UK and overseas, to support these findings and confirm the significance of women's community activities to both the shape of the rural community and the nature of rural women's gender roles. Middleton (1986), for example, provides a colourful illustration of the essential nature of women's labour in key village events, contrasting this with their exclusion from others (most notably the village cricket match). In another community-based study, Marshall (1999) considers the gender division of labour on the remote Canadian island of Grand Manan, New Brunswick, again identifying the importance of women's domestic work and its extension into essential and non-essential community activities. Teather's work on farm women's political and social networks in New Zealand, Australia and Canada situates women's involvement in rural organisations in the context of their broader community roles and the 'separate spheres' of male and female activities (see Teather, 1992, 1994). Murdoch and Marsden (1994) comment briefly on the centrality of women's roles to the creation and survival of the rural community in relation to their work on village development and planning in South-east England. As they conclude:

> There are a variety of social service organisations in the village . . . These services are primarily run by women . . . The view was expressed by certain respondents that it is the women of the village that really keep the village institutions going and make Weston Turville into something resembling a community. (Murdoch and Marsden, 1994: 109)

While many studies, then, make reference to women's roles within the rural community, far fewer have attempted to interrogate further the extent and nature of those roles. Hughes' (1997a, 1997b) work is an exception, however, going into some depth in documenting women's attitudes towards their community role and variation between women. Her partly historical perspective is valuable in identifying change in women's responses to the community and in suggesting that, for some mainly younger women, involvement in rural community activities has to be negotiated in the context of the demands of paid work. This is a point to which we will return later in this section. Hughes' emphasis on the lived experiences of women in the context of community involvement also resonates with Lipiens' (2000a) argument, discussed earlier, that our understanding of the rural community needs to be informed by the lives of those who live in it.

My own (1997b) research on rural women's voluntary work attempts to provide more detail on the nature of women's community involvement. Using data drawn from three different case studies, I identify more precisely the nature of women's community roles and the time commitment such roles involved.[1] The research yielded some rich data on the wide range of tasks performed by women under the guise of 'community activities'. It also revealed the extent of many women's commitment to the community in terms of time. Table 4.1 summarises the sorts of tasks listed by women when asked what activities they undertook within the community.

Not surprisingly, the list of activities performed by the women I interviewed reflected, very strongly, an extension of their domestic roles. Caring activities, involving for the most part children and old people, constituted one of the major forms of community-based work. Tasks included permanent care of their own relatives but also organisation of a range of formal and informal activities from basic care to more social events. Activities for children, in addition to the provision of day care in play groups and mother-and-toddler clubs, also accounted for another large chunk of community work. Such activities were arranged in connection with a variety of groups and included school-based activities. It was interesting that the majority of mothers interviewed helped on a regular basis in the local primary school and indeed perceived such 'help' as an expected part of their community role.

What was also apparent from this research was the extremely gendered nature of this 'community work'. Again the view that this was women's work was deeply rooted in the views of those interviewed. Many of the activities mentioned (running the old people's social club or the toddlers

---

1 In this work I acknowledge the problems of identifying precisely the full extent of 'community work' and of calculating the time commitment of individuals. The data reflect the particular respondent's own interpretation of 'voluntary work'.

Table 4.1 Voluntary activities performed by women living in Wiltshire and Avon villages

Church-based activities	Work with elderly	School-based activities
Sunday school	elderly ladies group	assistance to disabled
church visits	boat trips for pensioners	hearing schoolchildren read
PCC	social club for elderly	general help at school
ladies organ circle	transporting elderly to village functions	playground helper
church socials		Parent–Teacher Association
cleaning		school governor

Health-based activities	General
disabled riding club	citizens' advice
	charity fundraising work
	village jumble sales
	village hall committee
	playing field committee

Source: Questionnaire surveys undertaken in 1990 and 1993/4[2]

group, for example) involved no men at all while others incorporated men's participation in very specific roles or circumstances. In one of the villages, for example, there was a tradition of men organising certain sporting events or building a stage for the Christmas pantomime. A few women commented on the lack of male involvement in community-based work, drawing attention in particular to the selective nature of men's participation. Most women, however, echoed the sorts of views found by Hughes; she quotes one of her interviewees as follows:

> 'women have to persuade their husbands that they must join in . . . some men are very good but a lot of men won't unless they are pushed to do it. Things like that don't come easily to men . . . They are not good at community things like women are'. (Hughes, 1997b: 178)

Other studies have found that men's involvement in community activities favours positions (and activities) of power (see Seymour and Short, 1994; Stebbing, 1984). Men can be found on the parish council or parochial

2 Questionnaire surveys of women in connection with Rural Development Commission (RDC) funded research on women and rural employment (see Little *et al.*, 1991) and Polytechnic Central Funding Council research on women and rural lifestyles (see Little and Austin, 1996).

church council, for example, or running the village hall committee – posts that are perhaps more formal and visible. Men are less likely to 'muck in' and lend a hand running a pensioners' group or organising the village fête. This self-reinforcing situation stems from the same traditional expectation of rural gender roles as assigns to women the majority of the tasks in the village community; men talk of feeling excluded from all but a limited number of community activities. It is, however, too simplistic to dismiss men's non-involvement as a reflection of the lack of status and power accruing from the majority of village-based tasks. Women argue that their community role allows them considerable power and status in the village and that these roles need to be protected from the influence of men. As Hughes (1997b: 180) observes: 'women who helped out were held in very high regard and were well respected'. Clearly, there are some complex power relations operating between community members and while these are certainly fluid and will vary from place to place, they do incorporate some interesting questions for the operation of rural gender relations.

Running through the studies on gender and community roles is a widespread acceptance of the division of labour amongst members of the community. While clearly, as Hughes (1997a) notes, roles might be changing, the essential assumption that women 'run' the community remains intact at the local level. None of the women I talked to in the course of researching gender roles in the rural community expressed any dissatisfaction with their involvement in the community, save some who regretted that, due to other requirements on their time, they were unable to give as much time to community work as they would have liked. Few questioned the value of their contribution nor the demands that it made upon them. Elsewhere I have looked at the potential conflicts surrounding the replacement of formal (often state-provided) services, such as nurseries or day care for the elderly, by 'voluntary work'. Interestingly, as I discuss in more detail in that work, this conflict was not raised by the women themselves (see Little, 1997b). Thus their participation in community work was not contested on the grounds of equity either between women and men in the community or regarding the differential access to resources of different communities.

Women's acceptance of their role in the community appeared to be underpinned by a wish for integration. This links very closely to ideas about the nature of community in rural areas as discussed in some depth in the following section. In terms of the practicalities of women's involvement, the desire to belong to the community, especially for recent in-migrants, encourages them to seek to play their part in the pattern of community life and to accept the roles assigned to them. As Hughes reports from one of her respondents in Ditton:

> there is a lot of pressure in the countryside to conform to what society wants of you. If you don't try and fit in with the village and what is organised you are not accepted . . . You have to work to prove yourself to feel as though you belong there. (1997b: 179)

Women I interviewed in Wiltshire and Avon (see Little, 1997a, 1997b) showed a similar wish to fit in to the community and saw participation in community work as part of the process of initiation into village life. Many expressed a certain gratitude that they were 'permitted' to get involved, seeing such involvement as a mark of their acceptance into village society. Some talked, for example, of 'the honour' of doing the church flowers or 'being allowed' to help with the fête. This, as discussed later, is linked to the cultural construction of rurality and community and draws on ideas surrounding the uniqueness of 'country people' and 'rural ways'. Women, then, could not be accepted as true 'countrywomen' until they had 'learnt' to take part in community events and assisted in the organisation of village-based activities.

Acceptance of their role did not mean that women did not feel *pressure* to participate but rather that they saw this pressure as understandable and, again, as part of what belonging to a community was about. Hughes (1997b) talks of the pressure surrounding, in particular, new migrants to rural areas to show a willingness to get involved in the community. Such an obligation is gendered in both its direction and outcome, affecting women far more than men in the majority of cases. The assumption of women's community role means that they are the ones targeted for help and they are the ones most likely to experience fall-out from an inability or unwillingness to get involved. The size of the community and the visibility of those who do not 'join in' serve to strengthen the links between community work and acceptance.

The question of change in the nature of women's contribution to the rural community is raised by some authors (see Dahlström, 1996; Hughes, 1997b; Little, 1997b). Rural community studies have long been concerned with the 'demise' of the rural community, and research on social change in rural areas has invariably reported on the perceived negative impact of population migration on the 'sense of community' within the village (see Little, 1986; Newby, 1979; Phillips, 1993). Related to the idea of the community in decline is the view that women are now less prepared to get involved in community activities than they were in the past; indeed this reluctance has been offered as one of the *causes* of community decline, particularly by more elderly rural residents. Hughes (1997b) notes the criticism implied by her respondents of those who move to the countryside to 'escape' the city with no intention of contributing to the community. Such behaviour is not seen as in keeping with 'country ways' and represents not only a decline in community but an urbanisation of the village itself. Increasing involvement in paid work is the main reason suggested for women's reduced commitment to community work – as Hughes (1997b) again argues, paid work not only restricts the time available for community work but takes women away from the village during the day, thereby shielding them from the needs and effects of community activities.

Some research, however, has concluded that community activities and women's involvement in them in particular have not actually declined significantly (see Little, 1997b; Rogers, 1987). Interviewing women in Avon I found few who felt that the community was less active than in the past. It may be that the relatively high turnover of people in the village meant that those I spoke to were unfamiliar with the community activities of the past, but the list of tasks still undertaken suggested a continuing and buoyant commitment to community work. Moreover, comparing the level of community involvement of women with and without paid work suggests that employment has relatively little effect on willingness to take part in community activities. It is difficult to generalise on the prioritisation of different forms of 'work' undertaken by rural women; in some cases women I spoke to claimed that a shortage of accessible employment (especially part-time) left them free to engage in community activities and indeed helped to keep them from getting bored. Others, however, spoke of fitting voluntary and community work around their paid jobs and even talked of 'not having time for a paid job'. To some extent there appeared to be pressure on those women in employment to make sure that their community responsibilities did not suffer and that they were seen to be 'pulling their weight'.

This section has outlined the gendered nature of community roles and highlighted the major contribution made by women to a range of essential and non-essential tasks within the village. The activities themselves but, more importantly, the willingness to take part are predicated on the strong beliefs that surround the meaning of the rural community and the gender roles embedded within it. In order to understand more about the detail of women's (and men's) involvement in the rural community we need to look more closely at the social and cultural constructions of community and rurality, including, as Lipiens (2000b) argued in her framework for community-based research, examining the meanings attached by different people within popular discourses and imaginings to 'rural community'. It is to these questions of cultural construction and representation that I turn in the next section.

Gender and the cultural construction of the rural community

As noted in Chapter 1, there is now a substantial volume of work on the cultural construction of rurality in contemporary capitalist societies. Such work has emphasised the importance of the notion of community to the ways in which we understand and make sense of 'the rural' (Bell, 1994; Cloke and Little, 1997; Crouch, 1992; Jones, 1995). A feeling of community, and the ability to foster and hold on to all that this entails, is perceived, particularly in Britain, as something intrinsically rural, a quality that divides rural from urban and one that can no longer be (if it ever was) found in the city. It is argued here that this construction of rurality underlies much that has been discussed so far concerning involvement in community activities and the gendered nature of that involvement. Not only is the community seen as an essential element of rural life but its continued existence is dependent on (and at the same time reinforces) traditional gender relations and an accepted gender division of labour.

The version of community encapsulated within the dominant construction of rurality is highly traditional, drawing on romantic notions firmly rooted in the past. The rural community of the 'rural idyll' has been described by various authors as timeless, close knit and harmonious (see Laing, 1992; Short, 1991; Williams, 1973). It is a place where, although social divisions may exist, people are friendly, relationships strong and lasting and villagers share a common view of the importance of rural life. Critical to these imaginings of the rural community is a feeling of authenticity – a belief in the rootedness, both practical and ideological, of the community. With this authenticity comes a sense of honesty and the idea that rural communities are built on (and inspire) simple, genuine relations. Importantly these social relations, and the strong hierarchy within which they are framed, are largely uncontested; firm in the belief that there is a 'natural order' within the rural community which is understood by all. As Short sums up:

> The countryside as contemporary myth is pictured as a less hurried lifestyle where people follow the seasons rather than the stock market, where they

> have more time for one another and exist in a more organic community where people have a place and an authentic role. The countryside has become a refuge from modernity. (1991: 34)

Bell (1994) sums up how such sentiments are articulated in the views of village people. Living in the village of Childerley represents, for Bell's interviewees:

> Quietness, a slower pace, smallness of scale, knowing everyone, helping others, tradition, refuge from the rat race, advantages for family and children, freedom from material competitiveness, religious morality. (1994: 93)

Its separateness from the urban is one of the key features of the community of the rural idyll. The rural community hangs on to values and attributes that have long been lost to the urban community. The advance of the city, the infiltration of people with urban views has become seen as the main threat to the survival of the traditional rural community. Any challenge to accepted practices and views of how things should be done in the rural community is likely to be attributed to the influence of the urban and to the perceived desire for city folk to 'interfere' with 'country ways'. This portrayal of the urban as a threat to the rural community has been very clearly exemplified in recent debates in the UK over the future of hunting. The rural lobby, characterised by the pressure group The Countryside Alliance, has interpreted the Government's bill to abolish hunting with hounds as an attack not only on the rural economy but on the traditions of rural life and the values and solidarity of the rural community. A typical response to what the bill was seen to stand for by 'rural people' is quoted by Woods (1998) from the *Daily Telegraph* newspaper:

> We cannot and will not stand by in silence and watch our countryside, our communities and our way of life destroyed forever by misguided urban political correctness. (1998: 8)

A whole set of issues is incorporated within the debate over hunting and attitudes to rural life, including questions of national identity, power relations and the governance of rural society. What is important here is the sense in which these debates rest on a view of the rural community which is essentially distinct, separate and under threat from the urban and one which constructs a sharp divide between urban and rural people.

Similar ideas of the threat to the rural community from the town have also been demonstrated in recent responses from rural people to the threat and reality of rural crime. At the time of writing (November 2000), the Home Office had just publicised a report in which it was confirmed that 'fear of crime' is increasing in rural Britain. These fears apparently relate to 'criminals' coming out from urban areas to the less heavily policed rural areas. 'Crime' (especially violent crime and crimes against property) is perceived by rural people as an urban problem and further evidence of the encroachment of urban behaviour and values as a destructive force on the countryside.

The emphasis placed by dominant constructions of rurality on the stability of social relations in the countryside and on the traditional and enduring nature of the rural way of life should not imply that the rural community is static and unchanging. Many academic studies have drawn attention to the extent of population change in contemporary rural areas and, in particular, to the social transformation of parts of the countryside as a result of global and local economic shifts and changing lifestyles (see Hoggart *et al.*, 1995; Marsden, 1998; Murdoch and Marsden, 1994). Clearly there are different people now living in rural areas in many capitalist societies – people who often have no direct experience of 'country life'. Change in the class structure of rural societies has been especially significant leading to, in some cases, a shift in the power base in village communities and the emergence of new rural elites (see Woods, 1997; and Murdoch and Marsden, 1994). What is interesting, however, is the extent to which even the most recent in-migrants invest in established ideas of rural community and a belief in the continuing relevance of what are seen as traditional rural values. Indeed, as I will go on to argue later, it is often the wish to retain the old rural way of life that attracts migrants into rural areas and while their presence may entail something of a re-invention of community, their motives are based on a strong presumption of the uniqueness and superiority of 'country ways'.

The enduring nature of traditional ideas of community is tied up with the representation of rurality in popular culture. Since the early work of Raymond Williams (1973), various authors have examined the way the rural community has been depicted in forms of art, literature and other forms of media from classic novels to present-day soap opera (see Brace, 1998, 1999; Laing, 1992) and even, as Houlton and Short (1995) point out, in children's games. Such cultural references reflect and reinforce the rural community as a strong, tight-knit and positive force and while more contemporary depictions have injected a degree of realism into community narratives, these tend to overlay rather than displace more traditional ideas of rural society. Recent work has drawn attention to the influence of these dominant representations of the rural community not only on the attitudes of those living in or moving to the countryside but also on the selling of rural areas as heritage sites and tourist destinations (see Gruffudd, 1994).

As has been alluded to earlier in this chapter and in the book, the key issue here is that dominant social and cultural constructions of the rural community sustain and are sustained by conventional gender roles and relations. Chapter 2 has discussed how the development of feminist geography from the study of gender roles to gender relations was reflected in rural research by a shift from the study of issues such as access, employment and service provision to an interest in the household division of labour and power relations within the family and community. This shift opened up debate on the construction of gender roles and, in attempting to explain the reasons behind gender difference and inequality in rural areas, work started to look at how women and men's lives were situated within dominant ways of understanding and imagining the rural community. More recently, work on the nature of rurality itself has helped to develop some of these ideas on the relationship between the way rurality is viewed and interpreted and expectations of gender roles.

The much-quoted and highly influential work of Davidoff *et al.* (1976) provided an early articulation of the links between gender experience and the social and cultural construction of rurality. Their claims that women were seen as the 'lynchpins' of rural society, holding together the rural family and community, have been taken up and developed by subsequent writers (see Hughes, 1997a, 1997b; Little, 1987, 1997a; Little and Austin, 1996; Stebbing, 1984; Teather, 1994). As such work has sought to emphasise, the strong link between women, the family and the rural community is both practical and ideological; it defines their responsibilities but also their status in rural society. The belief that women's 'natural' place is at the heart of the family and, as an extension of this, the community, incorporates strong moral ideas on the behaviour of women both within and outside the family.

While it may be argued that identity of women as mothers and the dominance of women as the main carer are hardly exclusive to rural society, the strength of traditional ideas concerning gender roles in relation to parenting and family life and the lack of (overt) contestation of these roles do seem to be clear features of contemporary rural life in developed countries. The rural community is seen (and by some celebrated) as providing the social, cultural (and even physical) environment for the reproduction of women's 'place' in the family and community as sustained by hegemonic gender relations. Thus the notion of 'country woman' encapsulates an adherence to and valuing of women's identities as primarily mothers and wives (see Little, 1997a; Hughes, 1997a). It is, moreover, the mutually reinforcing nature of traditional constructions of gender identities and rurality that is important. This relationship is apparent in research that has been conducted amongst rural women, as I go on to discuss below.

The particular relationship between gender identity and the rural community, while powerful, is also perceived to be under threat from the urban and more generally from modernity. 'Urban influences', especially the involvement of women in paid work, are seen as challenging the 'natural' role of women and mothers, a role that rural society has clung on to. The respondents Hughes (1997b) spoke to in her study of women's lives in the rural community of Ditton blamed 'urban attitudes' and the desire of younger women for employment for the perceived decline in community spirit. While my own research has tended to suggest that women in paid work are as likely to take part in community activities, there remains a concern more broadly amongst rural women that participation in paid work will place new demands – practical, social and ideological – on younger women and families. An editorial in *Country Life* provides an interesting illustration of the concern that 'urban values' and aspirations (including paid work) may be damaging to rural social relations and that while women's desire for 'a career' may be important, rural women may have 'other priorities' to fulfil. The article states:

> Most people who live in the countryside have to some degree chosen to live there, and appreciate that the benefits of living in a rural community depend on all its members participating in it in a way that becomes impossible if everybody (ie women!) chooses to work elsewhere. It is certainly common to find women working in jobs for which they are far too well-qualified, but they are likely to have decided that the rewards of where they live outweigh the attractions of better-paid urban work.

It goes on:

> [Women] contribute greatly to the life of rural communities by voluntary work from unpaid assistance at village schools to care for the elderly. Such work is for most women not a compromise but a conscious choice and a source of fulfilment. (Aslet, 1999: 99)

And commenting specifically on the Government's Rural Audit (The Rural Group of Labour MPs, 1999), in which lack of employment choice for women was identified as a key concern for future rural policy:

> Although it is undoubtedly regrettable that women who want a career feel that rural conservatism is a hindrance, the Government needs to ensure that providing encouragement for such women will not be interpreted as yet another example of *urban values and priorities being foisted on the countryside*. (Aslet, 1999: 99, my emphasis)

Recent research undertaken by rural geographers on the rural community more generally has highlighted the importance of ideas of community, stability and tradition to people's understanding of rurality by examining their motivations for moving to the countryside. Such work has confirmed the power and durability of the 'rural idyll' as a key influence on local and regional patterns of migration (see Boyle *et al.*, 1998). Halfacree (1994) cites the 'quality of the social environment' as a key factor in what he describes as the 'lure of the countryside' amongst migrants. He suggests from his research in rural parts of Lancaster and mid-Devon that the following seven features form important attractions to actual and potential migrants:

- The area allowed one to *escape* from the rat race and society in general.
- There was a *slower* pace of life in the area, with more time for people.
- The area had more *community* and identity, a sense of togetherness and less impersonality.
- It was an area of *less crime*, fewer social problems and less vandalism.
- The area's environment was better for *children's upbringing*.
- There were far fewer *non-white* people in the area.
- The area was characterised by *social quietude* and propriety.

Boyle *et al.* (1998) also critique a range of other studies of rural migration from different countries (including the USA, Australia, Finland and Sweden), concluding that the importance of the community to people's desire to move to the countryside is not a specifically British phenomenon.

Interrogating this notion of social environment further (and with reference to the particular concerns of this book), one of the major attractions of the rural community appears to be its suitability as a location for family life and especially for bringing up children. In our research in Avon, Tricia Austin and I examine references to the family in respondents' views of rurality and the village community (Little and Austin, 1996). A total of 34 per cent of respondents[3] made reference to the social environment

---

3 Drawn from the Avon research in which 71 questionnaires were completed by women living in a village south of Bristol.

of the rural community as the primary reason for their choosing to move to the village, with over one third of these making specific mention of the value of the environment for children. Amongst the responses were the following quotations:

> 'We wanted to bring our children up in the country.'

> 'Clean air, beautiful surroundings and a better environment to bring up children in.'

> 'We wanted to live in the country and enable our children to go to a small village school.'

Bell also notes the stress placed on children and family life as positive aspects of rural life. He writes:

> Many Childerleyans also talked about the countryside as a better place for family. The phrases 'better for the children' and 'good for the family' are conversational cowslips in the village. Like that famous English country flower, they are touchstones of what is right and good about country living. Parents often cite the cleanliness of the countryside as well as the sense of the countryside as a place of traditional family and religious values. (1994: 93)

Talking to women in my research about their involvement in paid work, there was a strong presumption that a job was very much a secondary concern. As I describe in Chapter 5, the majority of women with children had left employment completely when their children were small and while some spoke of the frustration they felt in not being able to access adequate part-time work, most claimed to have made a conscious choice to give up paid work and concentrate on being a mother. Indeed, like the *Country Life* article discussed above, most of the women I interviewed felt that the choices were straightforward, and the prioritisation of motherhood had been obvious in the decision to move to the countryside. The view was that if you were going to have a job or worry about your career, then you had no business moving to a village. Thus while prioritising children over employment is not specifically rural, the assumption that a particular lifestyle (and with it a particular set of gender identities) is tied in with a particular type of community is.

The research also showed, as in Hughes' study of Ditton, that mothering roles provided important entry into the village community for women, particularly when they first arrived. Forty-five per cent of the women I spoke to were involved in the village play group or mother-and-toddler scheme or helped out at the school, and indeed only four of the 22 mothers with children under 5 did not participate in the organisation or running of these events. The reverse was also apparent in that the perceived failure to fulfil their duties as mothers ensured some women were not accepted by the community. Although it seems rare, remarks were made about women not 'pulling their weight' as far as childcare responsibilities were concerned. Again, there was criticism of such women for choosing to live in a rural area and the frequently expressed view that in failing to get involved in children's activities, women were defeating the object of living in the village.

There was little sense of this emphasis on motherhood as the key component of women's rural gender identities being contested. Indeed, a few women spoke of the attraction of the role and of the pleasure they felt in being valued as mothers. Professional women in particular believed that the emphasis placed on children allowed them to immerse themselves in their children's lives without feeling a constant need to improve their careers. This, they stressed, was peculiar to rural communities and contrasted powerfully with their experiences and perceptions of urban society.

This valuing of motherhood as a powerful element of the social and cultural construction of rurality raises questions about the position of rural women who through either choice or circumstance are childless. Women choosing to live by themselves in rural areas are relatively rare – although the high numbers of elderly people mean that older women are often left living alone on the death of a partner – and rural communities are overwhelmingly dominated by families. In the Avon research, for example, just 5 per cent of women interviewed were not living in a nuclear family. Again there can be some suspicion surrounding the motives of younger and middle-aged single women deciding to live in the countryside – a feeling that there is nothing for them in the village where many of the activities revolve around children and families. As one woman put it:

> I think it would be strange to live here without the children. So many of the village events revolve around children.

Bell (1994) found a similar emphasis on, and expectation of, family living in his Childerley study. He writes:

> As a resident described one . . . couple who were not married, 'They're, shall we say, "partners". And that's not the village way.' Most residents apparently agree, and only thirteen of the village's 185 households are set up in other than traditionally accepted arrangements of married couples,

bachelors, spinsters (no young women live alone in the village), widows and widowers. (1994: 213)

More alarmingly, Watkins (1998) found in her study of Oxfordshire villages that residents found the presence of single rural women unsettling. They were viewed as slightly odd, potentially dangerous or simply downright predatory. Clearly the acceptance of individual single women is partly about the characteristics and views of particular people and communities. In general, however, the rural community rests very firmly on the active involvement in and support for a culture of traditional family life. As noted in Chapter 1, any variation of this can seem inappropriate, threatening and very much 'out of place'.

Gender, community and village institutions

While there is considerable evidence of women's active involvement in community activities and organisations, the nature of their influence over village institutions is, as briefly indicated above, less clear-cut. As has been noted elsewhere (see Bell, 1994; Little, 1997b; Murdoch and Marsden, 1994; Stebbing, 1984), the gender division of labour in relation to community work is highly conventional. Just as women carry out the majority of activities associated with children, men tend to dominate in the sporting events. Women undertake the caring work and organise social functions while men perform construction (or destruction) work. This division of labour is also reflected in the membership and status of the various village organisations, with certain organisations being prioritised in, for example, the use of particular communal facilities and spaces or the amount of publicity given to their meetings or events. As Murdoch and Marsden note:

> Many men, especially those commuting to work elsewhere, tend to be away from the village for significant periods. In the village their activities seem to be mainly dominated by sport. A majority of men sit on the Amenity Committee, and there are golf and squash, sailing, sports and fishing clubs. Men are also active in the horticultural society. (1994: 109)

It has been suggested that village institutions can be a powerful influence on the reinforcement of conventional rural gender relations and on the traditional nature of rural women's gender identity. Frequently cited in this respect is the Women's Institute (WI) (see Hughes, 1996; Stebbing, 1984). Research has drawn attention to the continuing emphasis on 'domestic skills' of this important rural organisation, arguing that while it claims to be moving with the times and attempting to shake off its old image of 'Jam and Jerusalem', it is still dominated by an interest in home-making and domesticity. It is argued that the emphasis placed by the WI in its meetings on crafts and cooking encourages a view of rural women that is limited and limiting. While clearly such activities should be valued and the high levels of skills routinely achieved by women appreciated, the fact that the WI appears to promote such skills above all others is seen as reinforcing women's traditional gender role.

Hughes (1997a) makes the point that cooking skills, in particular, as promoted by the WI, are interpreted as rural. 'True' country women, so some of the respondents in Hughes' research argue, are able to cook well and have no truck with 'convenience foods' (see also Cavanagh, 1999). The ability to produce wholesome food, from basic natural ingredients, is seen as a rural skill and evidence of real rural authenticity. Again, the loss of the skill to produce 'real' food is seen as the influence of the urban over the rural and another change perpetrated by urban 'ex-migrants'.

> Although Jenny [a respondent] argues that images of rural women baking their own bread and churning their butter went out with the Ark, she suggested that real countrywomen would not dream of using convenience foods. Incoming women, in particular, defined 'real' rural women as possessing a rural background and being able to cook, bake and generally be domestic. (Hughes, 1997a: 132)

Other authors are more positive about the role of traditional rural organisations such as the WI in accommodating change in women's gender identities. Teather (1994) argues that while the WI and the Australian equivalent, the Country Woman's Association, continue to espouse traditional ideologies and more conservative outlooks, they are nevertheless 'useful precedents and even models' for the more radical organisations that have formed recently. These institutions, she suggests,

> became, and remain, formidable and effective lobbyists. They strove for what is, arguably, the necessary first stage in women's emancipation: the facilities that help them carry out their reproductive role efficiently, with confidence and without jeopardising their health. Furthermore, these early, gendered networks, and the spaces owned and managed by women, permitted new forms of awareness to develop elsewhere. (Teather, 1994: 45–6)

The more recent rural women's networks Teather identifies are, she argues, more radical in contesting the status and roles of women and in challenging established gender relations in rural (mainly agricultural) communities of Australia and Canada. In Britain a number of new women's organisations have also emerged in recent years – mainly organised around women's role in agriculture. While they may be somewhat more radical than the WI and less accepting of the exclusivity of women's traditional gender identities, their goal is the support and valorisation of women's central role in farm businesses rather than a direct attack on dominant constructions of gender identity or on the power relations through which these are reinforced.

In terms of other village organisations, some research has suggested that gender divisions exist not only in participation but also in the occupation of key positions (Little, 1986; Woods, 1997). Men continue to be more dominant in positions of power than women and more visible in the 'flagship' community institutions. This theme is taken up in Chapter 6 in the discussion of the gendered nature of power in rural communities and, specifically, the empowerment of women through the organisation of what have been termed new forms of rural governance. The encouragement of more active participation of the community in its own regulation (especially the support of 'partnership' approaches) may offer important opportunities for change in the involvement of all members of the rural community in local institutions of all kinds.

The gendered nature of community spaces

Although there are now a number of studies on the gendered experience of the rural community, very little attention has been given within such studies to the use of space and, in particular, to whether men and women have different spatial geographies of rurality. This is surprising given the academic and policy-related interest in the gendered use of space in urban areas. One explanation for this neglect may be that the space of the village is seen as 'natural', having evolved slowly in relation to the functions of the village – it is not perceived as having been *designed* in the way that urban space has been. A further reason may be that the study of the gendered experiences of urban space has frequently been motivated by issues of fear and safety. Again, these are not issues seen as relevant to rural spaces. Whatever the reasons, this lack of attention needs to be addressed – particularly, as pointed out in Chapter 2, as feminist work has stressed the importance of the embodied use of space to understandings of gender identity.

It can be argued that the powerful gender division of labour associated with community tasks assigns certain spaces, permanently or temporarily, as primarily male or female. The village school, the school gates or the village shop, for example, are women's space, at least on a day-to-day basis. The pub, the sports field (especially during a match) are male. Middleton (1986) talks of the 'public space' of the village as male (much as public space of urban areas is male). Perhaps more than in urban areas, however, the male domination and control of rural space is mediated in relation to particular times, circumstances and activities. Thus, the village hall may become women's space during the daytime meeting of the mothers and toddlers, but revert to being male space for the parish council meeting, or the more formal gathering of village action groups. Similarly, the village sports field becomes a male space for matches but can become women's space with the village fête.

Middleton also argues from her research on men's space and women's space in a Yorkshire village that men occupy public spaces of the village for recreation to a greater extent than women. She argues that women's use of spaces such as the village hall comes often in the course of their voluntary work. Exceptions obviously include the use of the village hall for social activities such as the WI meetings. Men, however, use spaces such as the pub exclusively for recreation. In one of the Avon villages I studied, the Village Club provided space for (mainly local) men to drink on a daily basis. Once a month, however, there was a 'social' which partners and wives attended. There was no reason why women should be restricted to drinking in the Club on these nights but it was accepted that for the rest of the time this was male space. For women to go for a drink on 'ordinary' nights would have been akin to men gatecrashing the WI, although no formal membership 'rules' existed.

Leyshon's (forthcoming) work on rural youth also looks at the gendered use of village spaces. His interviews with young people in Somerset reveal that the 'uncontrolled' public spaces of the village are used differently by boys and girls. While both sexes 'hang out' on the village playing field, the girls tend to gossip, smoke and simply escape the adult gaze; boys are more likely to play games of football, dismantle something (often an old car or bike) or undertake some sort of building project. The space thus becomes more incorporated into the activity rather than (in the case of girls) providing a location for it. Again Leyshon finds the use of the pub to be strongly gendered, with mainly boys getting involved in under-age drinking in the public space – presumably indicative of the greater tolerance of this form of transgression amongst male youth.

As Middleton (1986) again notes, the attitudes towards the gendered use of space are also influenced by whether or not girls and women are alone or accompanied. While many of the public spaces of the village are open to lone men, women using them have to be accompanied. Middleton describes her experiences living alone in a village in which she was conducting ethnographic fieldwork for research on rural women. She writes:

> In the second year of my fieldwork, having interviewed a number of women, I decided to drop in . . . at the pub at the end of the village . . . I discovered as a consequence that the boundaries between men's space and women's space were marked by a male clique at the pub with a continuum of controls ranging from chivalry and 'chatting up' through gossip and ridicule, to practical jokes and violence. (Middleton, 1986: 129)

She goes on to describe how other women in the village advised her to 'get herself a man' in order to prevent the attacks.

> There is an assumption that if a woman goes into the pub alone she does so in order to pick up a man. By entering a pub alone she declares herself to be 'fair game' and is open to a variety of controls from all directions. (Middleton, 1986: 127)

Women's fear of attack has been a frequently acknowledged influence on their use of space in urban areas. Again, this issue has remained largely unresearched in relation to the rural. There has been some work on women, safety and sport/recreation in the countryside as reported in Chapter 3, which has talked about the responses of women running or jogging in remoter rural areas and the strategies they used to avoid attack and reduce fear. There has, however, been little work on the dangers and fears of women using the countryside for less formal activities. Burgess (1995) does examine the responses of groups of women to walking in woodland but the woods concerned were in areas of the urban fringe and the women were walking in fairly large groups, not alone. There is a need for research on the feelings of women regarding the casual use of spaces in and around the village; which places they feel unsafe in and why.

Conclusion

This chapter has presented a very 'one-dimensional' interpretation of the rural community in drawing out some of the key associations with gender identity and difference. This is not to deny, of course, the existence of variations from this 'ideal type' of community and from the social relations framed within it. Later in the book, in the discussion of sexuality, I outline the existence of some alternative forms of rural community – forms in which the dominant relationships between gender and society that I have elaborated on here have been challenged or denied, and in which the assumptions surrounding rural gender identities have become destabilised.

But even in the 'typical'[4] rural communities of western capitalist societies, the patterns of gendered community expectations and experiences do not always conform to those outlined here. What I have done is to present what I (and some others) believe to be the main characteristics of the relationship between gender and the rural community. Such characteristics, as with those of gender identity itself, are fluid and changeable and of course they will not always be observed in the same way in every rural community. The individual histories, local politics and cultures of particular communities will influence social relations including the configuration of gender relations.

The chapter has touched on the issue of marginalisation and on the fact that women (and men) who do not conform to the dominant expectations surrounding rural gender identities may remain 'other' to mainstream village society and community. Yet, because of this 'otherness', the experiences of such people may go unnoticed; their real challenge to the accepted version of gender roles and relations within the rural community obscured by the continued dominance of conventional and traditional community forms.

Huge barriers stand in the way of the transformation of the relationship between gender and the rural community. Signs that shifts are occurring in, for example, participation in paid work amongst rural men and women or the involvement of women in local rural governance (see Chapters 5 and 6) do little, it would seem, to challenge the central place of the nuclear family and of women's domestic and community role. Rural communities, especially in the UK, are celebrated for their strong 'family values' and for the conservative nature of social relations. Such conventional household forms, and their associated gender identities, are maintained not only as

---

4 In using this term I recognise the problems of defining any rural community as 'typical'. While it could justifiably be argued that all communities are unique, the discussion presented in this chapter was based on a belief that it is possible to identify a number of dominant social characteristics of contemporary rural communities within capitalist economies and that such characteristics play a crucial role in the reproduction of rural gender relations and masculine and feminine identities.

offering the most efficient and appropriate ways of living in rural areas but as *natural*. The supposed natural order of gender relations and family life thus sits very comfortably with ideas surrounding the natural reproduction and operation of the rural community and landscape. Using the backdrop of gender and the rural community, Chapter 5 now goes on to look at one more specific element of the gendered countryside in the discussion of paid work.

5 | Gender, employment and the rural labour market

Introduction

The discussion of existing work on gender in rural geography in Chapter 2 identified employment as one area where there was a clear history of research interest. Studies of the gendered nature of the rural labour market, the characteristics of women's involvement in paid employment and the relationship between paid and unpaid work have been central to the development of a focus on women and on gender within the sub-discipline. Such work has contributed to a more informed understanding of both the detail of women's lives and of the characteristics of gender difference within rural society, and the theoretical debates surrounding gender relations and the social and economic relations around paid work. It is an area which has drawn heavily on feminist studies outside rural geography but one which has also developed in recognition of the specificity of rural social, cultural and economic change.

This chapter will provide an overview of existing research on gender divisions within the rural labour market. Making reference to a range of studies on rural employment, the chapter will identify the principal characteristics of male and female participation in paid work and discuss the major differences in the experiences and opportunities of men and women in the rural labour market. More than most of the chapters, the empirical detail of this chapter will mainly relate to *women's* experience. Looking at their involvement in the labour market and at the problems faced by women in accessing suitable paid work. The second part of the chapter will move on to an explanation of gender divisions within the rural labour market, locating the characteristics of women and men's employment within the broader context of rural gender relations. It will be shown how patterns of gender difference in terms of access to and experience in paid work relate to the operation of gender relations within rural society. Finally, the chapter will look at ways in which existing patterns of gender inequality within paid employment may be changing in rural areas in response to broader socio-economic shifts as well as specific policy initiatives. While the patterns of employment opportunity and experience discussed here are drawn from the UK, neither the trends observed in the availability of paid work nor the detail of rural labour market experiences are unique. Many of the characteristics and problems described affect women in rural areas of capitalist economies more generally and are part of the wider gendered experience of paid work within the rural labour market.

The rural labour market

In attempting to establish the precise extent and nature of involvement in paid work by either men or women we immediately encounter problems of definition. The difficulty of defining exactly what we mean by 'rural areas' and translating that into geographical spaces inevitably influences any attempt to identify patterns and measure changes. This issue has already been mentioned (see Chapter 1) in relation to other attempts to observe and discuss the economy and society of rural areas and will not be dwelt on here. Defining paid work itself can also be a problem. As is discussed below, the difficulty of separating production and reproduction has led to some interesting and lengthy debates over the nature of women's involvement in 'work' and in the labour market in rural areas, particularly in relation to agricultural work. Moreover, the temporary and casual nature of some types of paid employment (again, especially on the farm) can further complicate definitions and confuse analysis and understanding. Together these problems mean that any attempts to establish exact, quantifiable, patterns of gender difference within the rural labour market should be treated with care – especially those attempts which interrogate these patterns over broad areas.

Research using both primary and secondary data has revealed a number of major trends taking place within the rural labour market in the UK over the past thirty years. These trends include a massive decline in agricultural labour – particularly in terms of full-time permanent employment, as shown in Table 5.1. This decline has been documented, over time, by other authors (see Allanson and Whitby, 1996; Gardner, 1996) but it is worth

Table 5.1 Agricultural employment in England and Wales, 1891–1991

Year	Agricultural workforce (000s)	Percentage of workforce in agriculture
1891	1225	9.5
1911	1211	7.4
1931	1026	5.4
1951	921	4.6
1961	697	3.3
1971	521	2.4
1981	425	2.0
1991	379	1.8

Source: Allanson and Whitby (1996)

noting that the loss of jobs in agriculture continues, into the twenty-first century, to constitute a serious problem for rural labour markets and communities. As Gardner sums up:

> In the 1940s more than a quarter of the population of northern countries of what is now the European Community worked on the land; today the total is less than 7 per cent, and probably more than half of those who remain on the land also work part time, with the main part of their income arising from non-agricultural production. (1996: 72)

In keeping with well-established national and international patterns, rural areas in the UK have also seen a decline in the numbers of people employed in other primary industry and in manufacturing, with an increase in those working in the service sector (see Chapman *et al.*, 1998; Errington *et al.*, 1989; Townsend, 1991). During the 1980s considerable attention was focused on the relative success of rural areas in securing manufacturing employment. While losses occurred in this sector outside the main metropolitan centres, they were, it was argued, proportionately lower than in urban areas. This was commonly explained in terms of an 'urban to rural' shift in manufacturing employment in which the cheaper land and labour costs as well as reduced transport congestion enticed firms to relocate from the towns to the countryside (see Fothergill, *et al.*, 1985; Keeble, 1984). More recent research (see Keeble and Tyler, 1995) has argued that such a shift in the location of manufacturing continued to characterise labour markets into the 1990s.

In terms of broad employment trends, it has also been suggested (see Keeble and Tyler, 1995) that the service sector in the UK has experienced similar changes in terms of the location of employment growth, with rural labour markets benefiting from both the physical relocation of existing service-sector businesses and higher rates of new-firm foundation. In their research, Keeble and Tyler (1995) pay particular attention to the role of enterprising behaviour in the creation of new employment opportunities. They suggest that this has been a major factor in the growth of rural enterprises and played a key role in the continuation of the urban-to-rural shift. 'Enterprising behaviour' on the part of rural firms has depended very heavily on the ability of rural settlements to attract a relatively high proportion of actual or potential entrepreneurs due to the quality of the residential environments (see also Cloke and Thrift, 1990).

While these labour market trends are supported by data at the national and regional scales in the UK (see DOE/MAFF, 1995), care must be taken in identifying and interpreting their significance to rural areas at the local level. As will be shown below, the more detailed studies of specific localities and communities have frequently revealed variations from these broad-based national trends. The same warning must also be given in the interpretation of employment trends in relation to gender differences and although the key changes taking place in the rural labour market will inevitably include a gender dimension, the precise nature of gender difference will vary from place to place.

Notwithstanding these concerns about local specificity, structural changes in the rural labour market have resulted in a decline in the sorts of jobs traditionally done by men and an increase in those carried out by women. The loss of primary-sector employment, particularly agriculture, has had a serious impact on long-term trends in male employment in rural areas – research by Bennett et al. (2000) in the coalfield areas of England and Wales has identified persistent male unemployment in rural communities following the collapse of jobs in coal-mining, for example. Decline in primary industry has often had particularly catastrophic effects on localised workforces as the closure of a quarry or mine wipes out a major source of employment in areas where there are few alternatives; the coalfields research has identified associated problems of ill-health and poverty with the loss of local job opportunities. The increase in service-sector employment has generally not provided employment for men to replace that lost from primary and manufacturing industry, and new employment opportunities have been directed more towards women than men, as has been the pattern in the economy as a whole.

Just as important in terms of gender has been the change in the nature of employment itself that has occurred in rural (as in other) areas. Key changes here have been in the decline of full-time, permanent employment with a relative increase in temporary, casual paid work (see Chapman *et al.*, 1998). While clearly this is not an automatic replacement of 'men's jobs' with 'women's jobs', it does constitute a change in which the kinds of working conditions traditionally (for all sorts of reasons that will be discussed later) associated with women's employment have been given priority to those under which men have generally been employed. Women have largely been seen by employers as more flexible and adaptable workers – not necessarily as a result of their own choice, but nevertheless easier to 'hire and fire' than men and frequently more able and prepared to work in temporary situations. Thus they have been seen as more suitable in the context of the restructuring processes taking place in the labour market and the global move towards flexibility within the workplace. Work by Leach (1999) in Canada demonstrates the importance of such trends outside the UK; she argues that rural women are at the 'leading edge' of the shift from production to consumption and that it 're-emphasises the caring kinds of work traditionally associated with women' (Leach, 1999: 222).

While this move towards flexibility tends to be associated most directly with employment in the service sector – indeed, contributing to the very importance of this sector itself – other forms of employment have under-gone similar shifts. A very pertinent example in the rural context is that of agricultural employment. As noted, paid work in agriculture has been in decline in the UK and other western economies for at least forty years. Within this overall decline, there has been an added shift in the nature of the work that *is* still available in agriculture. Much agricultural employ-ment has changed from being full-time and 'permanent' to casual, seasonal and often part-time. Contracting-out specific tasks and hiring temporary help during periods of crisis have increasingly become part of the farmer's response to labour needs. In certain sectors (especially fruit, vegetable and flower picking) this has, again, been seen as a shift from men's work to women's work.

Another feature of the rural labour market is the emphasis that has developed on self-employment. Cloke *et al.* (1994) recorded an average rate of 29 per cent of respondents in self-employment in their study of rural lifestyles in England[1] – compared to a national rate of about 24 per cent . This 'average' for the study areas in Cloke *et al.*'s research, however,

---

1 Paul Cloke, Paul Milbourne and Chris Thomas undertook a major research project on lifestyles in rural England in the mid-1990s (and a supplementary study of rural Wales). They looked at a range of household and individual indicators across 12 rural areas. The results of this research have provided important evidence of levels of marginalisation and poverty in late twentieth-century rural communities.

masks figures as high as 44 per cent in some rural parts of the country. High rates of self-employment are reflected, not surprisingly, in the size of firms located in rural areas. Small businesses are particularly common in remoter rural regions such as the South-West – partly due to the reluctance of larger firms to locate in these areas and hence the lack of choice for those wishing to move to or remain in remoter regions. Self-employment tends to be more prominent amongst men; many women lack the capital and experience to set up their own business (although this may be changing for some groups of women). As discussed later in this chapter, small businesses are often characterised by low wages and generally offer less secure employment opportunities than larger companies.

The main characteristics of the rural labour market as revealed in existing UK-based studies thus seem to be a familiar decline in male employment in primary and manufacturing sectors – although opportunities in manufacturing in *some* rural areas have not shown the same rates of decline as in the UK generally – and an increase in service-sector employment, much of which is carried out by women. In terms of actual numbers of jobs affected by these changes (as opposed to the proportion of the labour market or of a particular sector), the picture is extremely difficult to clarify. As well as problems of definition referred to earlier, rural employment structures are complicated by commuting patterns – particularly those of professional male workers. Simply identifying the existing employment profiles of rural residents conveys little of the difficulties experienced by particular groups in gaining access to suitable employment in rural areas, nor of the conditions under which some workers are employed. Detailed research has shown wide variations in the employment structures of rural populations, with accessible localities demonstrating much higher levels of professional and managerial employment than the remoter parts of the country. In Cloke *et al.*'s (1994) work, for example, rates for professional and managerial employment were as high as 49.5 per cent of the workforce in West Sussex, while in Devon, North Yorkshire and Northumberland they were less than 20 per cent. There is also evidence to suggest a polarisation of employment patterns at the district level with relatively high numbers of both professional/managerial workers *and* unskilled/manual workers. A similar picture is found when it comes to the employment conditions under which rural people work. The rural labour market is renowned for offering relatively low rates of pay and poor working conditions, and yet such problems tend to be masked by the comparatively good pay and conditions of commuters.

Clearly there are implications in these arguments for the examination of gender divisions within the labour market. The employment patterns identified have, as noted, an important gender dimension. Our understanding of gender difference, however, will inevitably remain superficial if it relies on broad-based conventional labour market studies. It is therefore important that we move beyond such studies and focus specifically on gender inequalities and differences in their own right rather than as a by-product of other labour market changes. As has been argued elsewhere, central to such an analysis is the thorough examination of women's participation in the rural labour market.

Rural women's employment

As Chapter 1 has briefly noted, early detailed academic studies of women's employment in the rural areas of western capitalist economies focused on agriculture (see, for example, Gasson, 1980; Sachs, 1983; Whatmore, 1990). The aim of such research was generally to establish the role played by farmers' wives (and sometimes daughters) in the business of agricultural production. It sought to identify the extent and nature of women's labour on the farm, to clarify what tasks were generally performed by women and how these contributed to the profitability of the farm business. The work was inspired by a lack of detail on women's actual involvement in the farm and the belief (by some) of a corresponding and widespread underestimation of the scale and importance of women's labour. The work of farm wives, it was argued, was frequently taken for granted, generally unappreciated and almost always unpaid. Moreover, such work was not simply helpful in the running of the farm, it was essential.

Studies by Whatmore (1990), Gasson (1992) and Shortall (1992) amongst others have confirmed the key role performed by farmers' wives and by other women family members. In a study of 135 women from farms in Dorset and the London greenbelt area of the UK, Whatmore (1991), for example, found high numbers of farmers' wives to be involved in both the administrative and manual aspects of farm labour; 85 per cent of women surveyed dealt with enquiries, ran errands and were generally 'on call' for the farm business while 65 per cent undertook the farm bookkeeping and other paperwork. Similarly, 70 per cent of women were involved in the manual labour of the farm, 32 per cent as regular workers. Research has also explored the detail of this involvement, looking, for example, at the sorts of tasks most frequently undertaken by women and the levels of responsibility afforded to them. Such work, together with similar studies in the USA, Australia, New Zealand and Europe (see Almas and Haugen, 1991; Braithwaite, 1994; Rivers, 1992), has drawn important distinctions between different types of farm enterprise (distinguishing, in particular, the unique position of the family-run farm) and pointed to the changing nature of women's input into the farm business in response to changing levels of agricultural productivity (see Arkleton Trust, 1992).

Research on the labour of farmers' wives has also been extended to consider women's roles in farm diversification – their involvement in a variety of farm-based and off-farm enterprises beyond the scope of the traditional forms of agricultural production – and the centrality of such work to the profitability of the farm business. Most commonly cited has been women's involvement in bed-and-breakfast and tourism-related activities on the farm (see Bouquet, 1987), although women have also been shown to be engaged in a range of other money-making enterprises such as food processing, the production of clothing, farm shops and educational activities on the farm (Evans and Ilbery, 1996; Gasson and Winter, 1992). Recent consideration of the involvement of women in the production of farm-based food has shown how such work is frequently related to the agricultural climate; in the UK, for example, the 'crisis' in farm incomes in the late 1990s and into 2000 has led to the establishment of local 'farmers markets' as outlets for the sale (mainly by women) of farm-produced and packaged food. There is insufficient space here for a full discussion of the ideas contained in such research – it is important to note, however, that the academic interest in the lives and experiences of farm wives led, in addition to a better understanding of their role and importance on the farm, to important debates on the definition of women's work and on the relationship between productive and reproductive labour.

As well as providing evidence of what women do on farms, the farm wives research produced two important and far-reaching conclusions concerning the way women's labour has traditionally been viewed. First, it showed that women's work on the farm frequently went unrecognised (officially and unofficially); it was largely unpaid and simply expected, by both farmer and farmer's wife, as part of the 'way of life' on the farm – even where the tasks performed were clearly central to the farm business. Secondly, there was often little distinction between women's work on the farm and their domestic responsibilities. According to the farm-based studies, farmers' wives were nearly always responsible for the majority – if not all – of the domestic work of the farm household. They were found to be in charge of the cooking, cleaning, shopping and childcare duties associated with the household, and expected to perform these duties regardless of what other work they did on the farm – on either a routine or emergency basis.

This research on the undervalorisation of women's farm labour has made an important contribution to theoretical discussions of the division between productive and reproductive work. A body of predominantly Marxist-feminist writing emerged during the 1970s and 1980s which argued that women's reproductive work was essential to the maintenance of the household and the ability of household members to carry out waged labour. It was pointed out by some authors (e.g. Delphy, 1984; Redclift and Whatmore, 1990) that the family farm provided a very useful example of the problems of separating women's productive and reproductive work; it was claimed that the particular characteristics of agricultural labour as well as the spatial proximity of the farm business and the farm household reinforced the interconnectivity of the different spheres. Such work went on to suggest that the underestimation of women's productive work is essentially a gender issue and that any attempt to explain the lack of value associated with women's farm labour needs to consider the patriarchal control of the labour process and of the ownership of the means of production.

Writing more recently, Morris and Evans (forthcoming) draw attention to the fact that women's contribution to the productive work of the farm continues to be widely undervalued. These authors look at representations of women's roles in agriculture in the farming press and not the persistence of stereotypical images of women's (and men's) work. They write:

> The evidence suggests that both gender identities have proved remarkably resilient over the 20 years examined despite wider social change and restructuring within the agricultural industry. The investigation of 'Farmlife' has demonstrated that the British farming media does not challenge well-established gender identities in UK agriculture but continues to reinforce them in subtle ways. (Morris and Evans, 2000: 18)

A further element in the debate on the relationship between gender and agricultural labour concerns the role of technology. Early studies of farm women's agricultural contribution noted that, even where women were quite heavily involved in farm-based work, they were far less likely than men to be responsible for (or even use) large machinery and other items of new technology. It was suggested, in this respect, that the capitalisation of agricultural labour had reduced women's participation in certain sorts of work and encouraged the gendering of particular agricultural tasks (for example, combine harvesting and spraying were more generally carried out by men while work with small animals was mainly left to women). Recent research by Brandth (1995) has developed these arguments in a study of tractor use in agriculture. Looking at farm businesses in Sweden, Brandth notes the demarcation of machine work as 'men's work' and the continuation of the association of machinery with physical strength. She writes:

> Even though the tractor is becoming more comfortable and easier to operate, making differences between women and men in muscular strength less relevant to occupational success, the absence of women and feminine imagery/metaphors is conspicuous. Despite the fact that the observed changes in masculinity are in the direction of a more feminised model, the images tell us that farm machinery is still an important masculine area. (Brandth, 1995: 132)

This work provides a valuable insight into ideas about the cultural construction of masculinity and femininity in the context of the rural labour market and demonstrates the application of new theoretical concepts to the understanding of agricultural work, as has been shown in Chapter 3.

Despite the empirical and theoretical progress made in the study of farm women, research on the employment of rural women outside agriculture was slow to emerge. It was not until the late 1980s and 1990s that work on the job opportunities and experiences of rural women more generally became a focus of research in rural geography. Following this rather slow start there have now been a number of detailed pieces of research in the UK, North America, Australia and parts of Europe (see, for example, Dempsey, 1992; Hogbacka, 1995; Leach, 1999) on various aspects of women's paid work and on the particular experiences of different groups of women within the rural labour process. It is difficult to generalise from the results of this work since one of the main arguments for carrying out detailed local studies is that each rural area is unique and that the experiences of women will differ from area to area. Despite a reluctance to generalise, however, it is possible to identify a number of key themes that have emerged from the research. In discussing these themes the following section will focus mainly on the UK – some detail on other parts of the world and how patterns here compare to the dominant trends identified will, however, be included.

While, as discussed earlier, some rural areas in the UK and elsewhere have seen an increase in the kinds of jobs generally performed by women, some studies have pointed to relatively low levels of employment traditionally amongst rural women. For example, a Rural Development Commission (RDC) study of women's full-time employment undertaken in 1990 in three rural areas of the UK (see Little *et al.*, 1991) found involvement

Table 5.2 Women in full-time and part-time employment in rural case-study areas[2]

Case-study area	% women in full-time employment	% women in part-time employment
Wiltshire/Cornwall	54	58
Avon	60	75
National (UK)	71	42

of women in paid work in each of the areas to be lower than national levels (see Table 5.2). Varied pictures emerge from other research, however: Cloke *et al.* (1994), for example, found that in some areas economic activity rates for women exceeded national averages while in others they lagged behind.

More consistent is the emphasis which occurs in rural areas on part-time employment. In the RDC research, for example, 58 per cent of women worked on a part-time basis. Similarly, in a later study by Little and Austin in two villages in Avon, UK, rates of part-time employment amongst women as high as 75 per cent were found. A further feature identified in this research was the relatively high proportion of casual and temporary jobs that were being performed by rural women.

Leach's work in North Wellington County in rural Canada also found that women's employment frequently revolved around part-time and temporary jobs. The restructuring of the local economy here had resulted in the replacement of full-time, 'permanent' jobs with part-time, temporary, 'flexible' work. Leach cites examples from her study of women failing to find full-time work and having to rely on a combination of part-time jobs.

The stress on part-time and casual employment inevitably raises questions about the quality of women's jobs and the conditions under which they are employed. The links between part-time employment and poor pay and working conditions have been discussed in the literature (see, for example, Walby, 1988, 1997) and it has been shown that jobs that are either part-time or temporary (or both) tend to be characterised by a lack of formal employment contracts and entitlement to benefits such as paid

---

2 The Wiltshire and Cornwall case study was undertaken as part of research for the RDC on women's employment in rural areas (see Little *et al.*, 1991) and the Avon case study was part of a research project conducted at the University of the West of England with Tricia Austin (see Little and Austin, 1996). The national figures are from the 1991 census.

Table 5.3 Rural women's conditions of employment

Case-study area	Holiday pay (%)	Sick pay (%)	Employment contract (%)
Wiltshire/Cornwall	52	46	58
Avon	72	69	83
National	88	67	Unknown

holiday and sick leave, as well as poorer wages and opportunities for promotion. There are exceptions – for example many areas of the public sector (e.g. teaching or local authority work) where conditions are the same for full-time and part-time workers and pay is on a related scale. Part-time and temporary employment often carry other associations, even if these are not reflected in the actual working conditions; it is often assumed, for example, that those working part-time are not interested in career opportunities but are working simply for the financial gain, or that they are not concerned with training and promotion.

Research has found the pay and conditions of women working in the rural labour market to be generally poorer than in urban areas. The extent to which this is a function of the higher degree of part-time and temporary work is difficult to say, although it appears that some full-time employment in rural areas is also characterised by insecurity and low wages. Table 5.3 presents a comparison of employment conditions, based on holiday pay, sick pay and contracts, between rural women (surveyed as part of the RDC research) and national statistics.

The case studies of rural women's employment have found it difficult to identify broad patterns in rates of pay. Different methods of payment coupled with a reluctance on behalf of respondents to divulge their income mean information may be rather unreliable. Research by Shucksmith and others (see Chapman *et al.*, 1998) on national employment characteristics for households in urban and rural parts of Britain found a mixed picture when they looked at wage rates. Most significantly, they found that women in rural areas were more likely to experience *persistent* low pay than either their urban counterparts or rural men. In addition, it seemed that women were less likely than men to experience upward mobility in their wage rates.

Anecdotal evidence from case studies of women's employment suggests that in some rural areas wages may be significantly lower than for equivalent work in urban areas. Comments from women interviewed in the RDC research, for example, included:

'In hotel and catering pay is not good for what you have to do. Mostly the work is part time so they don't have to provide meals or breaks. Women and girls get a bad deal as they don't have the experience or nerve to ask for their rights. In lots of cases women are only working because their husband's wages are so low.' (woman respondent, Cornwall)

'Living costs are high, especially transport. Wages are lower – one job working in a bank in P...... was offering £5,500 for a typist. This would earn probably £8,500 a year up country.' (woman respondent, Cornwall)

In discussing low wages, many women we spoke to talked about the 'benefits trap' – a phrase coined in the UK to refer to a situation where people are worse off by taking a paid job because the wages they earn mean that they no longer qualify for state benefits. In many cases the wages earned are not sufficient to offset this loss of benefits, however, and so even where people can find a paid job they may not be able to afford to take it up. In our research a significant number of women claimed that the cost of rural childcare and travelling, coupled with the low wages and loss of benefits, meant that paid work was not an option. As one respondent summed up:

'Low pay here means that by the time you have paid for child minding its hardly worth it. I was offered part time teaching but it did not make financial sense once travelling and child care costs were taken into account.' (woman respondent, Wiltshire)

The benefits trap is particularly problematic for women who are single mothers. One respondent in Cornwall told us that she would rather have a paid job than live on benefits but she is a single parent and would need to find a relatively well-paid job to compensate for the loss of Family Income Support as well as meet current mortgage repayments and pay for childcare. Her lack of professional qualifications made it unlikely that she would be able to find such a job.

Leach (1999) explores the problem of low wages in her Canadian study of rural women's employment. She demonstrates the gap between full- and part-time employment, noting that the average hourly wage of women working full-time in 1996 was $11 while those working part-time earned just $7.05 an hour. As she sums up:

By 1996, the difference between average full-time and part-time wages for women had widened. This seems to be attributable to a polarization between those (few) women who were willing to seek full time jobs at some distance from their communities of residence, and those compelled (through their domestic responsibilities) to accept part-time jobs locally. (Leach, 1999: 217)

Linked to the issue of low pay is the fact that many of the jobs performed by women in some rural areas tend to be repetitive, boring and un-skilled.[3] Rural women are less likely to work in professional and managerial jobs than urban women (or their husbands); in the RDC research we found extremely low numbers of women working in these kinds of jobs (less than 1 per cent over the whole sample). In more accessible areas the availability of skilled and professional work was reflected in the greater representation of such work amongst employed women, but in remoter areas the picture was very different showing a concentration of women in semi-skilled and unskilled occupations (see Little *et al.*, 1991).

What comes through strongly from the various studies of women's employment in rural areas is the lack of choice available within the labour market. While case studies have suggested that there are reasonable levels of satisfaction amongst women in terms of the jobs they do, it is clear from more detailed discussion that this satisfaction is tempered by a belief in the absence of alternatives. Women questioned during the course of research (see Little and Austin, 1996) spoke of being lucky to have a job at all, while employers talked of the high numbers of women applying for posts available – particularly where these offered relatively reasonable pay, flexibility and good working conditions. Linked to this lack of choice, evidence suggests that many rural women who do have a paid job are working in areas or at levels that do not relate to their qualifications, training or previous experience. For example, the Avon research under-taken in 1993 by Little and Austin revealed that 52 per cent of employed women were working in jobs which did not use their qualifications or training. The majority of women interviewed as part of the research said that they did not have a career plan (many, for example, had left work when pregnant) and were working for money and to get out of the house rather than with any thoughts of developing or continuing a career in that particular area of employment.

When asked about job satisfaction, the replies of women generally seemed tempered by a recognition of the limited range of alternatives. As

---

3 According to conventional interpretations of employment 'skills' which frequently undervalue caring or domestic work and reflect masculine values surrounding work practices.

noted, having a job at all was, for some, a bonus in an area of poor availability and choice. Clearly this recognition of the small pool of local job opportunities makes it difficult for women to push for improvements in wages and working conditions and leaves them vulnerable to exploitation by employers. Those women who did reveal dissatisfaction with their current employment complained of low wages, stress and a lack of job satisfaction. By far the most common problems, however, were the inconvenient hours that some were required to work and the lack of mental stimulation and opportunities for progression.

As I have argued elsewhere (see Little, 1997a), this gap between women's employment experience and qualifications and the reality of their involvement in paid work, while important to our awareness of the availability of employment opportunities in the rural labour market, is not simply about the supply of work. It incorporates important and complex relationships between the rural labour market, the household and the community. The job choices made by rural women are clearly influenced by the availability of paid work but also relate to their gender identities and to the gender relations within which their employment participation is negotiated and played out. These are important issues to which I will return below.

Returning briefly to the dynamics of the rural labour market and to employment opportunity, while most studies have not attempted to distinguish between different groups of rural women, there is some evidence to suggest that the problems of accessing suitable employment are felt more acutely by certain sections of the rural population. The research in Avon concluded that younger women, especially school-leavers, experienced particular difficulty in finding permanent paid work. The only options locally were casual work (unskilled and very poorly paid) in a yoghurt factory, mushroom farm or egg-packing factory, while to find more skilled work or jobs involving an element of training women had to move away or commute to urban centres. The conclusion that young women were particularly disadvantaged by the absence of choice in the rural labour market simply emerged from a more general study of women's employment. Clearly the precise nature of the problems faced by young women in finding paid work, or indeed the problems of other groups within the rural labour force, can only be fully explored through more directed research aimed specifically at uncovering the differing experiences of women based on age, class, sexuality etc.

Having introduced some of the key issues surrounding rural women's employment opportunities and experiences, I will now turn to a consideration of how such characteristics may be explained. In so doing I will look beyond the detail of the rural labour market to consider how paid work fits within a broader configuration and negotiation of household roles and gender relations. I will also consider the role of paid work in the formation of rural women's gender identities and in so doing draw on some of the theoretical issues raised in Chapter 4 concerning the construction and representation of women's identities in the rural community.

Women's employment and the rural household

The circumstances under which women become involved in paid work are as relevant (or possibly more relevant) to their employment experiences as the configuration of the rural labour market. The domestic and community responsibilities of women, the activities and interests of other members of the household, the attitudes of the family, the community and employers, for example, all act to influence the practical and ideological contexts of women's employment. The jobs women do as well as their career histories are a result of the complex and dynamic interrelationship between various elements of their lives and of the numerous constraining factors that operate within the individual, family and community. Gender relations are clearly central to the existence and interaction of the different influences and to the choices made around paid work, and any attempt to understand employment participation within the rural community must take into account the broader relationship between men and women. Thus, as I have argued elsewhere:

> Looking at the intersection of labour market, household and community rests on a detailed understanding of the broader lifestyle and life cycle strategies negotiated within the household, the assumptions and expectations of gender relations in the community and the ways in which the operation of both the household and the community construct, reflect and reproduce gender identities. (Little, 1997a: 147)

The assertion that participation in paid work is bound up in the operation of gender relations is hardly original. Studies conducted within a range of disciplines have articulated the ways in which gender relations in the home and the wider community have influenced the take-up of paid work and the sorts of opportunities available to both men and women (see, for example, Hanson and Pratt, 1995; McDowell, 1999; Massey, 1994; Walby, 1997). Such studies have been located in major theoretical debates surrounding the conceptualisation of women's paid work and, in particular, its relationship to the reproductive labour of the household (see Redclift, 1985; Walby, 1986). In addition, a substantial literature on the influence of gender relations on the experiences of men and women *within* the workplace has added to our understanding of the role of gender relations at both the empirical and theoretical levels (McDowell and Court, 1994; Walby, 1990).

As with other areas of work on gender, the vast majority of studies on employment and gender relations have been conducted in an urban context. Some rural research has, however, begun to apply arguments raised to the study of rural employment and while many of these arguments (especially those concerning theoretical perspectives) are not spatially dependent, it is important that they are mobilised in the explanation of rural as well as urban employment patterns. Moreover, recent interest in the cultural construction of rurality and its relevance to the attitudes and beliefs surrounding rural social structures and practices has highlighted the spatial specificity of certain aspects of gender relations and gender identities (see, for example, Hughes, 1997a; Little and Austin, 1996). It is likely that these aspects, while not distinctly rural in themselves, will take a particular form in rural society and exert a specific influence on employment practices and their explanation.

Chapter 4 has stressed the centrality of rural women's domestic role to their gender identity, arguing that women are seen – by themselves and others – as first and foremost wives and mothers. The traditional interpretation of women's role has, it may be argued, very important implications, both practical and ideological, for their involvement in and attitudes towards paid employment. Paid work is frequently seen as something that is peripheral to rural women's lives – it may be necessary to finance expenditure within the household or it may be a luxury of women keen to have interests beyond their family and the village – but it is rarely seen as part of their primary role. The initial research on women's employment in rural areas looked to explain their experiences, particularly the problems faced in accessing paid work, by reference to the practicalities of countryside living. The lack of public transport together with the poor provision of services such as childcare and health facilities were seen as barriers to women's participation in paid work and as potentially reinforcing traditional gender divisions in the labour market.

While research revealed many women to be frustrated by the practical difficulties surrounding their entry into and performance within the rural labour market, there seemed to be little renegotiation of employment or domestic responsibilities at the household level as a result. Women responded to practical difficulties in a variety of ways: they organised complex childcare arrangements to cover periods when they were at work, they took lifts from friends and called on family and friends to help out with children at times of 'emergency'. Most frequently, however, they reorganised their own paid work – changing shifts, working late or part-time and, in some cases, giving up employment altogether where no solution to childcare or travel needs could be found. Halliday (1997), in a study of childcare in rural Devon, cites cases where women had established complicated and often exhausting routines to allow them to take up paid work without neglecting their childcare and other domestic responsibilities.

'Jane', for example, one of the interviewees in Halliday's survey, was a qualified nurse with three school-aged children. She had several jobs, including working in an old people's home two nights a week and various cleaning jobs round the village. In school holidays she gave up the cleaning, looking after other people's children instead. She continued to work in the old people's home as this was night work that she could do when her husband was at home. Jane would like to resume a professional nursing career but feels that this is not possible at the moment as she believes her children would suffer if both she and her husband had full-time jobs. This pattern of women working night shifts or weekends so as to be available for their children when their husbands were at work was not uncommon in Halliday's research. It was exacerbated by the long hours worked (or spent travelling to work) by their husbands, whose employment needs always seemed to predominate.

Similar findings are presented by Hughes in her research in the villages of 'Ditton' and 'Langeley' on the Welsh borders (see Hughes, 1996, 1997a). She writes:

> Paid employment undertaken by village women tended to be informal and part-time . . . It was fitted around their childcare commitments and domestic responsibilities, and tended to be near their homes, and certainly in the village. In arguing these points Rebecca highlighted her own situation: 'I was at home when I had my kids although I used to go to work in the pub at Langeley . . . I used to clean in the morning . . . but then I could take whoever was the baby with me.' (Hughes, 1997a: 129)

Similarly, in the Avon research (see Little, 1997a) mothers described to me how they juggled the demands of employment to fit in with childcare:

> 'The strain with younger children is always on the mother to provide the childcare. Before my present job (as a secretary) I worked as a cleaner and drove a fish van to fit in with school hours.'

'As I work nights my husband looks after the children. Then I sleep while they are at school.'

'If the children are at home either ill or holidays I don't go to work (except occasionally during the summer holidays when they either come with me or go to a friends to play).'

Although the rural environment was clearly seen as an important factor in the practical difficulties facing women in gaining access to employment, research also stressed that such difficulties were also related to (if not directly caused by) the operation of gender relations within the rural household. Indeed, it was argued by some (see, for example, Little, 1987) that the two could not be separated and that even practicalities such as an absence of public transport were essentially overlaid by the priorities and assumptions derived from gender relations. Feminist geographers argued that the patriarchal nature of the rural household ensured that the domestic, reproductive duties were seen as primarily women's responsibility and that involvement in waged work by women was dependent on these responsibilities being carried out. Reorganisation of the gender division of labour or of assumptions surrounding, for example, access to the family car or childcare priorities were rare since this then challenged the balance of patriarchal gender relations and the powerful position of men within the family.

My own research has confirmed the dominance of conventional household structures and gender relations in rural areas, as described in Chapter 4 (see Little, 1997a, 1997b; Little and Austin, 1996). The implications of these traditional structures have been evident in a range of choices made in relation to employment within the rural household. In the majority of cases the male of the household was the main wage-earner and the primacy of his employment was not questioned. Just 8 per cent of women who had a paid job were the main income earner in the household[4] – in all other cases they were joint or (more usually) secondary wage-earners. The primary place of men's employment in the households interviewed was reflected in the life-cycle decisions that had been made in the family. For example, 23 per cent of those questioned stated that their move to the village had been prompted by a change in their husband's job – although, interestingly, in no cases was the woman's own job given as a reason for household relocation. In general these life-cycle decisions that

---

4 This was mainly because there were no men in the household.

had been taken in respect to moving to the village included not only the husband's job but starting a family and buying a bigger house.

Other choices concerning women's employment also reflect the gender relations of the family and the traditional nature of roles within the rural household. Rural women's career histories tend to be rather fragmented, generally showing little progression, and while many have had careers when younger, few appear to have maintained those careers through other changes in the household. The Avon research revealed, for example, that relatively few women had returned to paid work after the birth of their first child and while many hoped to start work again in the future, most were doubtful of being able to pick up their careers where they had left off. Seventy-six per cent of mothers questioned in the research had had a paid job before their children were born but less than one third of these returned to the *same* job when they went back into the labour market (see Little, 1997a). Women most likely to go back to work for the same employer following maternity leave were the professional women in quite senior positions (such as a bank manager and a GP). In no cases did women mention any changes to their husband's job resulting from the birth of a child.

Of course, this is not a distinctly *rural* pattern and is a feature of women's employment more generally. A number of points did, however, stand out which appeared to link the characteristics of the gender division of labour specifically to features of rural society and the rural household. Many of the women interviewed described their decision to stay at home with their child/children as part of a set of family life-cycle decisions which included living in a rural community. In some cases this coincided with a change of job for their husband or a promotion which would give them the opportunity to move house and pursue a 'more rural' lifestyle. As has been argued elsewhere, the cultural construction of rurality – the assumptions surrounding the sort of lifestyle and behaviour appropriate in a rural community – provides a powerful influence on values and decisions of rural residents. In the case of women's employment, dominant constructions of rurality suggested that women's 'proper place' was at the heart of the family and community and not chasing career opportunities. As explained in Chapter 4, women with young children appeared to have accepted that 'for now' their role was at home with their children and that if they had wanted a career they would not have moved to the countryside.

While some did comment on the importance of paid work as a form of mental stimulation and satisfaction, their comments did not reveal any great dissatisfaction with their roles – rather, a resigned acknowledgement that opportunities for combining paid work with being a mother were bound to present more problems in rural areas. As one woman said:

> 'Being a mum is much harder than being a soft furnisher. Being a mum is very fulfilling but you become isolated from the outside world unless you work as well. It's very exhausting living here and doing both but for me it's the only way to keep my brain alive.'

Disruption to the dominant pattern of men as primary workers and women in the home came, in the Avon research, only in the case of very few professional couples (where men and women had jobs of equal socio-economic status, for example a husband and wife who were both dentists) or where men had been unexpectedly made redundant. This second reason was not common but had recently become an issue in the area due to the downturn in the economy and the effect this was having for some forms of professional employment – for example solicitors and architects. Anecdotal stories were presented during the course of the research about couples where women had gone back into paid work following their husbands' loss of employment. In such cases considerable strain was seen to have been placed on the family – interestingly, this was felt to be most acute in two cases where women had started their own businesses which had become highly profitable. Again these kinds of responses are not specifically rural but were seen to have an added resonance in that they disrupted the more traditional gender division of labour within the household and as a result often threatened the lifestyle ambitions and expectations of the families concerned.

There is some indication from the various (limited) studies undertaken on rural women's employment that attitudes towards the appropriateness of women's involvement in paid work do vary. Hughes (1997a) reports on the highly traditional views of the more elderly women interviewed as part of her research; one, 'Phyllis', states:

> 'I think women's place was traditionally in the home and I am old-fashioned probably but still think so . . . that is what has gone wrong with our society. Women are going out to work and they are not doing their job at home looking after the children and the home.' (Hughes, 1997a: 130)

Younger women, Hughes argues, share such views to some extent but appear to be more flexible in attitudes towards paid work. As other research has indicated, many of the younger women in Hughes' study did have a job. Generally, however, that job was part-time and, crucially, had not required any changes to the women's domestic responsibilities. In other words, more younger women in rural areas may be involved in employment than in the past but such work fitted in around their domestic tasks – especially childcare. Hughes found (as we did in Avon) that incomers to the village, while generally younger and more likely to have been in paid work, had sometimes moved to the village to 'escape' the pressures of modern living – one of which was the pursuit of a career. Moving to a village, it appears, gave women far greater scope for following a peaceful and quiet domestic existence without feeling they should be out in the world of employment.

The rural labour market is still seen as limited even by younger women seeking employment. While attitudes may be shifting amongst rural women, there is still a belief that opportunities for employment are much greater in urban areas. Dahlström's research on young men and women in the rural periphery of Norway reveals that women there are increasingly looking to leave their home village in order to find paid work. Moving away from the rural area in search of employment is seen by young women as a way of resisting the traditional gender identities that exist in the Norwegian periphery. Men, on the other hand, appear to find more acceptable models of masculine identity in the countryside, supported by employment opportunities. For them there is less pressure to move away (see Dahlström, 1996).

Finally, it is important to note that also relevant to the gender division of labour in rural areas is the sort of jobs done by women. The fact that these are likely to be part-time, or casual, unskilled and poorly paid, has already been mentioned. Also noted was the redundancy, for many women, of previous employment skills and qualifications. In addition, research has suggested that women's employment in rural areas tends to be an extension of their domestic role. Jobs in the service sector – especially where they are tourism related – often involve domestic work such as cooking and cleaning. Cleaning for other village families is a common form of employment amongst rural women – particularly the older women, although younger women also spoke of taking on cleaning jobs as a way of ensuring flexibility in hours and amounts of work. Even where women's employment does not conform to these characteristics there is an assumption (amongst other villagers and employers) that the form of women's employment, by its nature, is not important or critical. Women spoke to me in the Avon study about employers assuming that their job was just a stopgap and that they would be unreliable. One woman who worked freelance for a computing company talked of her employers holding back work in the school holidays on the assumption that she wouldn't be able to do it because of childcare. Thus even where women (and families) go to considerable lengths to arrange childcare to allow for employment responsibilities, the image prevails of rural women as primarily mothers whose commitment to paid work will always be partial and contribution unreliable.

Women's employment and policy

This next section is concerned with employment policy and looks at the ways in which policy has influenced the gender division of waged work within the rural labour market. Emphasis will be placed on the direct relationship between gender and employment policy in an examination of those policies and initiatives containing a specific gender element. It is recognised, of course, that policies not designed to address gender issues may have gendered implications and outcomes, but to extend the analysis to include such policies would be impossible here. It is also the case that policy outside the field of employment will influence people's capacity to take part in paid work, often in a gendered way. Again it is not possible to cover such policies to any significant extent in this chapter.

Perhaps the first thing to note in the analysis of rural employment policy is the relative scarcity of initiatives aimed at issues of gender inequality. While some recognition is given to gender divisions in the labour market and to the particular problems faced by women in accessing employment by policy-makers in, for example, Rural Development Area (RDA) programmes, government policy documents such as the Rural White Paper and county- or district-based rural strategies, there are few examples of concrete policy initiatives to address such issues. A review of RDA programmes in the late 1980s (see Little, 1991) revealed an almost complete absence of policies designed to address gender inequality. The review showed that while 12 of the 20 RDA strategies analysed acknowledged gender divisions within the local labour market and recorded the low activity rates of women in their area as a 'problem', only three strategies followed this up with a related policy objective. Moreover, as the review further demonstrated, none of the RDA strategies contained any initiatives aimed at meeting women's needs.

Another recent rural regeneration initiative in England, Rural Challenge, designed to promote economic and social development within the same RDAs through job creation and other schemes, demonstrates similar trends as far as gender inequality is concerned. The Rural Challenge scheme makes awards of up to £1 million to six projects each year from bids put forward by partnerships within the RDAs of England (see Little *et al.*, 1998 for more details). Of the 23 projects funded during the first four rounds of competition (one was funded but subsequently folded) two made reference to the particular needs of women within the rural job market. One of these projects (Stainforth, near Doncaster in northern England) had introduced a training scheme to help women to return to the labour market while the second had funded the creation of childcare places to facilitate women's involvement in paid work. In both cases these initiatives formed supplementary parts of the Rural Challenge scheme in that area rather than the central element.

Recent attempts to explain (or at least contextualise) the absence of women's initiatives within rural policy have focused on changes in rural governance and on the policy process itself. This work is discussed in detail in Chapter 6 but it is worth making a few points here where such work contributes directly to an understanding of employment policy. It has been argued, for example, that the shift from a system of government to governance in rural areas has had important implications for the way in which decisions surrounding economic development and employment provision are made. Attempts to broaden participation in the mechanisms and practices of policy have encouraged the development of new networks of interests and decision-makers. It has been suggested that such networks are frequently dominated by business interests and local key actors – many of whom are men. Moreover, it has also been claimed that the types of mechanisms for allocating money for employment have been designed to encourage more aggressive competition and entrepreneurialism amongst businesses and policy-makers. The Rural Challenge initiative has been identified as typical of the kinds of directions and priorities reflected in new regulatory mechanisms. Pertinent here is the view (expressed by those working in both the urban and the rural sphere – see Tickell and Peck, 1996; Little and Jones, 2000) that such kinds of mechanisms represent very male-oriented styles of policy-making.

Clearly, it would be overly simplistic to suggest that rural employment policy lacked initiatives aimed at addressing gender inequality because economic regeneration policies were dominated by male interests and masculine styles of decision-making. Placing employment policy within a broader context of governance is, however, important even if care must be taken in assuming a direct association. It has been recognised for some time that the rural state is characterised by more traditional attitudes towards gender issues than its urban counterpart (again this is expanded on in Chapter 6). The broader policy environment would suggest a more traditional emphasis as far as gender initiatives are concerned and the indication from the examination of the recent policy initiatives such as Rural Challenge is that the shift towards new forms of rural governance will not see a major alteration in this emphasis.

As has been seen throughout this chapter, women's and men's employment is fundamentally tied in to assumptions surrounding gender responsibilities and the nature of gender relations in the household. Women's employment opportunities are, in the majority of cases, dependent not only on the supply of jobs in the local labour market but also on the availability and accessibility of a range of services and facilities which allow them to carry out the other aspects of their gendered role. Thus their participation in employment will be affected by, and possibly even dependent on, policies relating to health, transport, retail provision and so on. Of course, men's employment will also be influenced by changes in service provision. It is, however, likely that, because of their more traditional reproductive household role, and because of the priorities that are applied to employment in most rural families, women's employment will be more directly and immediately affected.

While the current direction of economic regeneration policy and rural regulation may not promise a dramatic increase in employment opportunities for women or a change in the circumstances which allow their involvement in paid work, there are some examples of individual or localised initiatives that have helped women to secure a paid job. Under the European Union-funded LEADER II initiative (Liaison Entre Action de Développement de l'Economie Rurale) a range of community-based schemes has assisted women in returning to the labour market. In Devon, in south-west England, for example, a training programme called 'Stepping Stones to Employment', funded by LEADER II, has been targeted specifically at women not currently in paid work and has aimed to provide training in personal development and confidence-building as well as in basic IT skills. The scheme provided training and guidance to a total of 38 women over a six-month period from September 1998 to March 1999.

Such community-based initiatives may not result in any radical change in basic employment opportunities in rural areas; they do, however, start to challenge existing barriers to women's involvement in paid work and unsettle some basic assumptions about the gendered nature of rural employment. The Stepping Stones to Employment initiative allowed links to be formed between different local agencies involved in training and childcare. It helped to establish community forums in which interested groups could come together, assess the needs of the local people and start to work towards integrated solutions. While the number of direct employment opportunities created by the initiative was small (three part-time jobs), it identified the employment and training needs of local women and put into place some strategies for helping women to access IT facilities while still carrying out childcare and domestic responsibilities.

In the absence of coherent or integrated policies aimed at addressing the employment needs of women in rural areas, there are some examples of individual companies and businesses developing their own initiatives. In Devon, for example, a company called OSM which produces magazines for parents of primary school-age children, employs only mothers and operates a 'flexible' attitude towards women, allowing them to take time off to care for children when they are ill and attend school functions. Initiatives such as these have persuaded the local authority to introduce a 'charter' which local companies can adopt to encourage them to take sympathetic attitudes towards employees' family demands. This and similar policies do little to challenge traditional gender roles but enable women to undertake both domestic and paid work. In some ways it could be seen as further cementing the mothering role of women, although for many mothers the opportunities to participate in paid work are so few that the kind of initiative operated by OSM is seen, by them, as a positive measure.

Conclusion

This chapter has examined the gendered nature of the rural labour market, focusing, in particular, on the employment of women. Building on the discussion in Chapter 4 on gendered identities in the rural community, the chapter has shown how women's experiences in paid work are not simply a function of the rural labour market but are strongly influenced by gender relations within the household and the construction of masculine and feminine identities. The emphasis on women's family role constrains their ability to participate in the rural labour market and shapes not only their access to paid work but also the employment conditions under which they work. Rural women are rarely seen as 'career women' and while more younger women have paid jobs than in the past, involvement in the labour market does not replace women's domestic role. In other words, paid work undertaken by most rural women is in addition to, rather than instead of their more traditional roles as wives and mothers.

Although shifts taking place within the rural labour market have opened up greater opportunities for women, many jobs remain poorly paid and insecure. Low wages generally within the rural labour market mean that there is significant poverty amongst some rural households. Research has shown that while rural areas in the 1990s continue to exhibit higher wage rates overall than urban areas, the figures are skewed by the very high incomes of some professional workers (see Chapman *et al.*, 1998). Certain groups – in particular the elderly, women and single parents – have been recorded as vulnerable to persistently low income levels while overall, as research by Shucksmith *et al.* concluded (see Chapman *et al.*, 1998: 40), a third of individuals in rural areas have experienced at least 'one spell of low income over the last five years'. For single parents (most of whom are women) the route out of poverty and low wages remains particularly problematic due to the high costs of childcare and transport in rural areas.

In middle-class households the low wages earned by many women who do have a paid job may not be as critical in terms of the overall income levels within the household. They do, however, reflect a lack of choice within the rural labour market and the persistence of traditional attitudes to women's involvement in waged work. While in some areas of the rural economy – notably farming – women have fought hard for their roles to be valued, there is also concern that the recognition of women as members of the labour market would constitute an attack on their domestic role. Many of the women I spoke to during the course of my research felt that their involvement in casual or part-time work or in voluntary activities left their status as wives and mothers (and thus 'true' country women) intact. Many would have liked more choice in the job market and some complained of low wages, poor hours or boring jobs, but it also suited them to keep their paid jobs as secondary to their main family caring roles. As long as women remain primarily judged by and valued for their domestic roles, little pressure can be brought to bear on employers to address the issues of employment pay and conditions or on policy-makers to try to improve the availability of employment or the provision of training.

6 | Power, gender and rural governance

Introduction

One of the main lines of argument running through this book is that gender relations are a form of power relations fundamentally located in and reflected by the relationship between women and men throughout all areas of life. So far attention has focused on the personal, and largely informal, expressions of that power relation. Where power in a more formal sense has been referred to (for example in the gendered nature of policies for rural employment), it has been in the context of specific topics rather than in respect to the broader operation of social and economic relations within the rural community. In this chapter, however, we turn to look more directly at the formal arena of power in rural areas. Discussion here centres on the policy-making process in an examination of the gendered operation and conception of power at a range of different levels. It looks at both overarching questions of governance and citizenship in the direction of rural regulation as well as more specific, local-level analyses of the practice and outcome of rural policy. In this chapter the main focus is again the UK, with the examples emanating from UK-based research. The national specificity of most of the agencies, organisations and structures discussed makes it perhaps more problematic to generalise to other countries. Moreover, the cultures of governance and the roles of various groups involved in the policy process mean that many of the points made are highly specific to the local context. Some of the observations concerning more general directions of power and the experiences of women may, however, provide useful pointers for the discussion of gendered power relations in other countries.

As with many of the debates elsewhere in the book, one of the key issues in this chapter concerns the appropriateness of the category 'rural'. Again, the intention is not to assert the uniqueness of the relationship between governance, political power and gender in rural societies and economies. Rather, it is to highlight both the contribution of a rural perspective to debates surrounding power and the gendered nature of the policy process, and the need to apply arguments raised elsewhere in the examination of governance and decision-making to an understanding of rural power relations. It is argued here that the different agencies and institutions operating within the rural policy process at all levels from the central state (e.g. the Ministry of Agriculture, Fisheries and Food [MAFF] and the Countryside Agency) to the parish (e.g. parish councils) provide some justification for a specifically rural focus, as do the existence of some powerful and spatially constructed rural elites (e.g. the Countryside Alliance and the National Farmers Union). More important, however, is the influence of particular beliefs and representations about rural decision-making and politics that underpins a rural perspective on the debate. Thus the rationale for this chapter rests to a large extent on the argument that the evolution and operation of rural power and decision-making, and hence the relationship between rural power and gender relations, are intricately bound up in the social, cultural and economic constructions of rurality and their historical negotiation and contestation. The fact that all rural areas may not exactly share such histories is obviously important to the understanding of the particular experiences of any individual group or place. It does not, however, deny the value of making more general observations about rural power and policy and their relationship with rural gender as it occurs across a large part of rural Britain.

The chapter examines the gendered operation of power, as stated, at a number of different policy levels. It also combines theoretical and empirical material. In the discussion of broad questions of governance and regulation, in particular, the chapter draws on recent theoretical debate from the urban politics literature. In so doing, the value of using work outside 'the rural' in the analysis and understanding of rural patterns and processes is reinforced. In addition, the contribution (hopefully) of such 'rural'-based work will be apparent generally within governance debates. It is with these wider theoretical questions of governance and politics that the chapter begins, highlighting the relevance of shifts in the nature of government to the relationship between gender and the policy process, before moving on to a more detailed examination of policy and implementation at the local level.

Gender and the rural state

Both the development of the theory of governance and the policy process and attempts to apply such theory to the understanding of local expressions of decision-making have been disproportionately concentrated (and constrained) within the urban sphere. Studies of the nature of government, the direction and form of policy and the role of different institutions, agencies and actors as well as debates around the conceptualisation of such issues, have been predominantly located in and applied to the regulation of urban areas. Rural areas have either been subsumed within or ignored by these studies and consequently an understanding of *rural* governance and policy has been largely dependent on the small quantity of work originating from, and specific to, the rural sphere. The reluctance, already noted, of rural geographers to actively engage with theory in discussions of social and economic change during the 1970s and early 1980s distanced them further from a major body of innovative work on the nature of government and the policy process, and helped dissuade them from seeking to explore issues of power and decision-making (a notable exception being work by Newby *et al.*, 1978 on the power of the farming industry – although this also avoided broader analyses of government and rural policy).

It was only really with the application of political economic approaches to rural studies that an interest in the analysis of questions of political power, government and policy-making, and in the theoretical debates surrounding these issues, began to emerge within rural research (see Cloke and Little, 1990; Hanrahan and Cloke, 1983; Marsden *et al.*, 1986; Moseley, 1982; Urry, 1984). During the mid-1980s growing interest in the national and international forces behind rural restructuring and their political and economic contexts encouraged rural geographers to start to examine the construction and operation of power within a framework of political change. A need to understand the characteristics and implications of rural restructuring in a variety of contexts (from agriculture to town and country planning) led to work not only on the content and direction of rural policy but also on the national and local priorities, values and assumptions within which rural decision-making was conceived and implemented.

More recently, a new focus in urban geographical research on government and policy, emerging from the identification of what has been termed 'new governance', has again been slow to translate to the rural context. A huge theoretical and empirical literature has been amassed examining the basis of the shift from government to governance in the urban context and charting its implications for the practices, mechanisms and cultures of local and national policy-making (see Amin, 1995; Goodwin and Painter, 1996; Jessop, 1993; Lauria, 1997). This has not, however, been supported by rural-based research. Thus, Goodwin, writing in 1998, identifies what he terms the 'curious neglect' of work on governance by rural researchers. This he sees as particularly surprising given the major changes that have taken place as a result of national and international patterns of rural restructuring and broad-based shifts in the mode of economic and social regulation. Such changes have given rise to a range of new structures and organisations of governance which, as Goodwin (1998) notes, need to be examined by rural researchers in an effort to understand not only the detail of local policy-making but also the wider trends in the way rural economies and societies in the developed world are governed.

Despite the neglect identified by Goodwin, there have been some recent attempts by rural researchers to address the gap in research and to start to look at the changes taking place in the form and direction of governance. Work by, for example, Cloke and Goodwin (1992) and Marsden (1995) has examined the shifts taking place in the regulation of the countryside in relation to the restructuring of agricultural production and the emergence of what has been termed a 'differentiated countryside'. Such work has argued that different 'types' of rural area have experienced the processes of restructuring differently, as discussed in Chapter 1, and that, as such, they exhibit variable degrees and forms of regulation (Marsden, 1998). In other studies attention has focused more on specific aspects of rural governance (such as community participation – see Marsden and Murdoch, 1998) and on the local practices incorporated within the shift from government to governance, and how these have been articulated in particular communities (see Edwards and Woods, 2000; Murdoch and Abram, 1998; Ward and McNicholas, 1998). The relative infancy of this work, however, means that it has not yet penetrated widely into studies of rural economic and social change.

It is not surprising that within this general history of rather limited research and publication on rural governance and policy, work specifically looking at gender relations within the rural policy-making process is particularly lacking. While there has been recognition from (some of) those interested in urban politics of the significant gender implications of many of the issues and questions that have been the focus of research – and indeed attempts to examine such implications – the analysis of rural policy-making has not, by and large, been characterised by a similar recognition. Only recently have attempts been made to start to examine the relationship between gender and the rural policy process. As with urban-based work, this examination has focused mainly on establishing the role and experiences of women in various aspects of governance and decision-making. While limited, it does provide a starting-point for this chapter and helps in the identification of key issues and important questions for future study. What is argued here, however, is that an understanding of the implications of the formal policy process for rural gender relations (and for the everyday experiences of women and men within the rural community) does not rest entirely on the study of the workings of rural governance but can also draw on wider observations of the rural society and economy.

Representation and women's initiatives in central and local government

The most obvious place to start any discussion of gender and policy-making is with the representation of women in the political process. Various studies have outlined the changing distribution of political posts amongst men and women in the UK and elsewhere and, in particular, the growing involvement of women as actors in key positions of authority (see Coote and Pattullo, 1990; Watson, 1990). In the UK there has been a significant rise in the number of women members of parliament, for example, in the last decade – although the low base from which this rise began means there is still an overwhelming numerical bias in favour of men. In the 1997 UK general election 120 women were elected as MPs (an increase of 100 per cent over the previous parliament), but this only constituted just over 18 per cent of all MPs. The point has often been made, in examining gender and representation within the political system of the UK, that while men tend to predominate at the national level, especially as MPs, women are disproportionately involved in local politics – especially in informal activity. Below I consider the representation of women in other positions of power, such as leading roles of key agencies and institutions, making the point that it is not only through occupying political posts that political power can be exerted.

As far as gender divisions and women's political representation in rural areas are concerned, the difficulty of separating 'urban' and 'rural' constituencies (due to their size and boundaries) makes it impossible to say with much precision whether women's political representation is better in the town or the countryside in the UK. A look at the list of sitting MPs, however, does indicate a concentration of female representatives in the metropolitan areas. At the local level, there tend to be fewer women elected as either district or county councillors, and while women do get involved in considerable numbers in parish councils, the leading roles in such organisations are more likely to be undertaken by men. In Devon, for example, of the 45 county councillors, only 11 are women and in a sample of parishes men consistently outnumbered women and occupied the formal positions. In the district of South Devon, only 4 of 48 parish councils have more women councilors than men and overall there are twice as many male councillors. Only 7 of the parish councils have women chairs.

Moving on from actual political office and the representation of women in the political system, research has also looked at gender in terms of the direction and content of policy. Work in the 1980s and early 1990s sought to highlight the continued inequality in national and local policy-making through, amongst other things, an analysis of what were termed 'women's initiatives' in local authorities. Halford (1989) identified formal structures ('women's committees') within county and district councils in England and Wales which, she claimed, were evidence of a local authority's willingness to take gender inequality seriously and to commit resources to addressing the unmet needs of women within their local areas. Women's committees and sub-committees were seen as raising the profile of women both outside and within the policy process (see Riley, 1990) and their formal constitution as highly positive at a time when women's involvement in political activity was still recognised to be largely confined to the informal sector. Generally their role was to ensure that women's needs were addressed (or at least considered) throughout the authority's activities. They often had budgets to initiate schemes and develop projects aimed at attempting to alleviate problems faced by women. It is not the intention here to discuss the nature of these projects nor to comment on their success or failure but simply to recognise women's initiatives generally as structures employed to draw attention to and to prioritise issues of gender inequality and equal opportunity.

Halford's research found that despite a high political profile, relatively few formal women's committees were initiated. In 1986, for example, only 8 per cent of all local authorities in England, Scotland and Wales had introduced either a full or a sub-committee (or comparable initiative) aimed specifically at women's needs. The important point here is that those formal women's initiatives that *had* been introduced were highly spatially concentrated. The greatest number had been created within the metropolitan districts – 30 per cent, for example, were in the Greater London area – while the vast majority of all women's initiatives occurred in urban authorities.

There are many possible explanations for the absence of formal local authority women's initiatives outside the urban areas – and several are significant in terms of explaining wider characteristics of the gendered nature of power and the policy process in rural communities. Some explanations concern the particular nature of urban politics in the 1980s and the relationship between the so-called 'New Urban Left' and the feminist movement. In the early 1980s new social movements, mainly mobilised around welfare and consumption issues, were creating an active and buoyant local politics, supported in many urban areas by the local state (see Boddy and Fudge, 1984; Duncan and Goodwin, 1988). The fact that such movements were clearly based on, or at least aligned to, party politics in an attack on the Thatcher government, served to reinforce an urban bias. Feminist politics developed as an important strand of the new urban politics and many of the women's initiatives that emerged at this time were a function of the role of feminist groups within the broader-based urban political movements. Although this relationship was frequently in negotiation and conflict, it did mark out women's initiatives in local government as very much part of the *urban* political agenda (see Little, 1994a).

Rural authorities were therefore often distanced not only physically but also ideologically from the political struggles of the early and mid-1980s. Their mainly Conservative or Liberal political orientation meant that rural local authorities were rarely sympathetic to the broad directions of the urban local state, while at the grass-roots level many of those living in rural areas did not share the aspirations of those active in urban-based politics. There was a strong feeling that the majority of the issues at the centre of urban political activity were of little relevance to those living in the countryside. An examination of women's initiatives in planning, for example, undertaken in the early 1990s revealed that the absence of positive action on behalf of many rural authorities reflected a widely held belief in such areas that the concerns behind such initiatives were 'not relevant' to rural areas. As one planning officer stated:

> There is no specific attention given to women's issues. The department is small and the district is rural. (Local authority planning officer, quoted in Little, 1994b: 181)

There was a sense, amongst other responses, of authorities not wanting to 'rock the boat' as far as traditional power relations were concerned and an underlying argument that women's initiatives were 'not needed' because there wasn't 'a problem'. Thus the research demonstrated a strong belief that women's needs were urban based in relation both to the problems experienced and to the potential local state responses. Again, such views are clearly of relevance to an understanding of the broader relationship between policy, power and gender relations within the rural community.

The changing form and position of local authority women's initiatives have been documented in some depth elsewhere (see, for example, Brownill and Halford, 1990; Little, 1994a). Through the late 1980s and early 1990s women's initiatives in the local state were generally shifting and unstable. By 1993 numbers had declined still further and only 17 local authorities still had a women's committee as part of the formal structure of the council. In many cases changing priorities and local political concerns had ensured the downgrading of women's issues in the context of local authority activity – commonly through the amalgamation of so-called 'equalities issues' and an assertion that it was the council's policy to insist on equal opportunities in all areas of its work. This move reflects a decline of the women's movement as a force behind local political activity. Feminist politics in Britain had lost impetus in the 1990s in the absence of major focuses of resistance such as the Greenham Common peace camp and the miners' strike of the mid-1980s and as a result of the fragmentation of the women's movement over internal racial and class divisions. But as far as formal local state activity was concerned, the fate of women's initiatives was also a function of a shift in the very nature, direction and purpose of government in its replacement, as noted above, with what has been broadly termed 'governance'. Before turning to look at the implications of this shift to a system of governance it is important to consider something of the effects of the absence, as documented, of formal structures in the public sector for addressing women's needs in rural areas. In so doing, the intention is not to take the analysis right down to the resulting inequalities experienced by women (since these are the subject of other chapters), but to see how an absence of women's initiatives is reflected in (the absence of) rural policy.

Women's initiatives and rural policy

Having established an apparent lack of interest in the establishment of formal structures for addressing women's needs in rural local authorities, it is unsurprising that policies aimed specifically at women or at gender inequality in rural areas are difficult to track down. While evidence from any general assessment of policy may be limited, studies of specific policy areas (for example planning and economic development) reveal such a clear absence of policy initiatives of this kind that it is difficult to believe that these findings would run counter to a more broad-based trend. For example, a study of policies aimed at reducing gender inequality and addressing women's needs in planning (Little, 1994b) found a very high degree of spatial concentration. Very few policies existed outside the major urban areas – and where they did, they tended to be one-off initiatives rather than part of a general strategy aimed at reducing women's needs. More specifically on the issue of safety, only 3 per cent of district councils in England had a planning policy aimed at women's safe use of the built and natural environment, while over 60 per cent of London boroughs had some such policy (see Little, 1994b for discussion). In Chapter 5 the absence of formal policies aimed at women's employment needs as part of the Rural Development Area Programmes devised by the Rural Development Commission (RDC) was noted. This again provides evidence from a specific policy area of the low profile given to women's needs in rural areas and is indicative of the view that gender inequality within the formal policy process is essentially an urban issue.

It would be wrong to suggest that this absence of policies aimed at women was a highly contentious or contested issue amongst rural dwellers. The traditional bent of the public sector in rural areas, referred to above, largely reflects, in the case of gender policies (and politics) at least, a shared wish (at least on the surface) to preserve the status quo in the formal arena of the local state. In other words the equation of 'women's issues' with urban areas is something that the rural community also appears to endorse. Rural areas in Britain have not been characterised by feminist politics or by the formation of women's groups other than in the form of more general activity or church-based groups such as the Women's Institute and Mothers' Union. While the WI as an institution has argued strongly in recent years to change its 'traditional' image in favour of one that is more contemporary and politically informed, it retains a strong apolitical stance. That is not to say that rural women have remained totally outside local political action – they have a long history of involvement in, for example, campaigns to save schools or other local services, or those in support of environmental or farming disputes – political battles that clearly impinge greatly on women's lives. Such involvement has, however, been mainly on the basis of single-issue campaigns and not generally aimed at improving the position of women *per se*. Indeed, many rural women would actively distance themselves from the idea that they were involved in a political act, especially one constructed around gender inequality, when supporting local campaigns.

The line between formal and informal political activity is, as is clear throughout this chapter, extremely important to the representation and the reality of women's involvement. In the discussion of gender and politics at the village level below further attention is given to women's role in decision-making within the community and to the relationship between the formal decision-making positions (on the parish council, the village hall committee etc.) and those associated with voluntary or community activity, particularly in terms of gender role assumptions, and to the appropriateness of women's political activity in rural spaces.

While the actual level of women's participation in formal politics as elected representatives is clearly important, this chapter is, as noted above, equally concerned with the broader decision-making framework within which gender relations in the formal policy process are played out. The actors involved are simply one part of the equation and any attempt to understand the nature of rural politics and decision-making, particularly the implications of and for gender, must also consider the *way* the policy process operates, and the values and assumptions subsumed within it. Earlier it was suggested that this wider construction and operation of decision-making in rural areas has been influenced by the recent movement away from an old system of government to new forms of governance. It is this shift, and its implications for the gendered nature of policy, that this chapter will now examine.

New rural governance and gender relations

As yet, little research has been undertaken on the relationship between new forms of governance and gender relations in either the urban or the rural sphere. The limited work that has been done, together with the conclusions that can be drawn from more general studies of governance and policy, has highlighted various issues of direct relevance to this chapter. It is clear that while specific evidence may be limited, the sorts of changes documented in the direction, style and culture of governance do have profound implications for the priorities attached to gender issues and for the configuration and operation of power within the formal policy process.

Tickell and Peck (1996) have undertaken one of the few recent studies of gender relations in a new governance context. In their research they chart the changing involvement of women in local politics and discuss how this has been reflected in the gender division of 'top jobs' in the institutions of the local state. They argue that the gains made by women in the formal arena of local government following the introduction of women's committees and the higher profile of feminist politics were by and large short-lived, falling victim to changing power relations within the local state. Thus, they write:

> Just as women began to make inroads into structures of power in local government . . . local government itself came under a sustained attack. Initially, this was manifest as restrictions on raising and spending money, along with compulsory marketization of services . . . More recently, however, quangos and non-state bodies have become increasingly involved in local governance. (Tickell and Peck, 1996: 604)

Although local quangos are typically drawn from local business people and other central government appointees, the numerical representation of women in such organisations is, according to Tickell and Peck's research, relatively high with something like one-third of local quango directors (in 1994) being women. These authors argue, however, that this apparent representation of women is little more than a token gesture and in fact masks a more qualitative decline in the political role and influence of women. A closer look at the distribution of women within the new quangocracy reveals them to be largely concentrated within the welfare-orientated organisations and almost completely absent from those concerned with economic regeneration and employment. This is particularly significant in the context of major shifts in the balance of local state power in favour of economic interests. So much so that, as Tickell and Peck argue, whatever women's representation elsewhere, their absence from the new 'business politics' effectively ensures their 'pervasive exclusion' from local structures of power.

Using the case of Manchester, Tickell and Peck show how as part of the shift from government to governance there has been a clear privileging of economic over social interests and a rise in the importance of (economic) elite groups over those emerging through consensus. Such changes, they assert, have constituted a masculinisation of local governance – a process that affects not only the policy and direction of the local state but also the *ways* in which decisions are taken and the policy process managed. This is very aptly summed up as follows:

> There is some truth in the assertion that the world of local politics is becoming more like the world of business and, as a result, it is also becoming more of a man's world. This has implications not only for issues of 'internal' governance, such as the operation of committees and the mechanisms of representation, but it also matters for the overall structure of local governance and, ultimately, for the constitution of local political priorities. (Tickell and Peck, 1996: 607)

The Manchester research also shows how the field of economic development itself has become increasingly male in its outlook. This is explained partly through the incorporation of business elites into local economic policy with the growing influence of the private sector in local state economic development initiatives. 'Business culture' is seen as patriarchal and as reinforcing a gender division of labour in which male work is valued more highly than female work. Again, where women do have a presence in the institutions of business and/or economic development (in, for example, the Training and Enterprise Councils), their involvement tends to be in the areas of equal opportunities, special needs and community initiatives (Tickell and Peck, 1996).

Empirical evidence is provided by Tickell and Peck to illustrate their arguments concerning the increasingly male nature of local governance. They cite a number of examples which they claim demonstrate how the discourses and *modus operandi* of the local state have become masculinised. This, they contest, may not be a deliberate (or conscious) attempt to exclude women on the part of government or business, but nevertheless serves to undermine women's participation and perpetuates a culture in which men are seen as the decision-makers and the powerful political actors in the city.

While clearly the local state and the operation of local economic development in Manchester constitutes a very different political and policy-making arena than would be found in the average rural area, there are important arguments made by Tickell and Peck, and by others on whose work they draw, that are applicable to the discussion of power and the policy process in the rural community – points, in particular, about the dominant interests and actors in policy-making and the gender implications of the emphasis, in local economic development strategies, on the entrepreneurial 'business' culture of local governance. Also very relevant here are the observations made about the masculine ways of managing local state decision-making and the patriarchal cultures emerging in working practices. The next part of this chapter will attempt to apply some of the main observations of Tickell and Peck to the rural context. As noted earlier, the lack of direct research on rural governance and gender makes this a difficult task and one that must frequently rely on indirect evidence and, at times, speculation. I draw here, however, on recent research on local economic regeneration in rural areas – research which did include a focus on gender and which, in the examination of rural economic policy, did encounter many of the issues concerning the emergence of governance and the changing culture of decision-making so critical to the understanding of gender within the contemporary policy process.

There is little doubt amongst those writing about rural governance that profound changes have taken place in the regulation of rural economies and spaces and that such changes conform, to some degree, to those identified as the so-called shift from government to governance by those writing from an urban perspective (see Goodwin, 1998; Little *et al.*, 1998; Marsden and Murdoch, 1998). These changes include the priorities adopted by the local state in rural areas, the range of organisations and agencies involved in the rural policy process and the ways in which the institutions and structures of the local state work. Authors such as Cloke and Goodwin (1992), Marsden (1998) and Woods (1998) have located changes in rural governance within a broader discussion of rural regulation and of the recent transformation of the rural economy and society. They show how the decline in agrarian economic and political power, together with the growing importance of a new set of competing rural discourses based around development, lifestyles and the environment, challenged the established way of doing things. As Goodwin writes:

> The 'ordered rule' was . . . beginning to change as the old landed elites operating through an established national and local government structure, came under political, cultural and economic pressure. (1998: 7)

Others have also argued that broad national shifts in, for example, the power and responsibilities of the public sector, the involvement of non-elected private-sector agencies in the provision of local services and the prioritisation of economic development in local regeneration all impinged on rural areas, further disrupting the traditional balance of power and operation of the policy process. The accepted institutional structure of local government in rural areas has been replaced by what Jessop (1995: 310) refers to as the 'tangled hierarchies' of governing; the new organisations, partnerships and coalitions through which power is mediated and applied.

The belief that rural policy-making has been influenced by these kinds of profound changes in the governance of rural areas is widely supported amongst rural researchers. There is also, however, agreement that the precise nature and implications of such changes will vary from place to place. The broad assertions concerning the characteristics of rural governance are mediated at the local level and the resulting changes to the direction, structures and practices of decision-making form a complicated and unique blend of national and regional processes and local social, cultural and political relations. The identification of the exact nature and implications of the shift to a system of governance for the rural policy process thus necessitates the collection of detailed, grounded information and its analysis within an understanding of local socio-economic characteristics. It was this mix of national and local information that our research on rural regeneration policy attempted to provide and which makes it useful, as noted above, in the detailing of some of the changes associated with new governance in the rural arena.

Rural regeneration, Rural Challenge and new rural governance

The particular form of rural regeneration policy examined in the research discussed here was an initiative entitled Rural Challenge. While not a major initiative in terms of either funding or numbers of awards, it is, we argue in the research (see Little *et al.*, 1998; Jones and Little, 2000), important here in that it exhibits, in its conception, formulation and implementation, a number of the trends and values associated with the shift from government to governance. It represents, moreover, an early example of the articulation and formalisation of such trends within a specific example of rural policy. Rural Challenge (RC) was a regeneration initiative that operated between 1994 and 1997 within the Rural Development Areas of England.[1] It was devised and administered by the RDC but abandoned with the replacement of the RDC in 1999 and subsequent introduction of the Countryside Agency. The initiative consisted of six awards of up to £1 million to partnerships of public, private, voluntary and community groups for the purposes of economic and social regeneration. Awards were competitive – local projects first bid in a county 'round' of the competition before being put forward by the county in a national round. Hence awards were spread, with no counties qualifying for more than one award a year.

It is apparent from just an outline of the basic terms of RC that it includes a number of 'new' directions/departures for rural policy and that in so doing it appears to take on some of the characteristics of new governance. The initiative constitutes the first application of direct 'staged' competition to the allocation of funding for rural regeneration. This form of competitive bidding became a hallmark of urban regeneration policy under the Thatcher governments of the 1980s. City Challenge (on which RC was modelled) – beloved of Michael Heseltine and supported

---

1 Areas designated as requiring particular attention due to the low levels of employment opportunity and economic activity and high incidence of poverty and deprivation amongst rural households.

by Conservative administrations – was designed to pit 'teams' of bidders against one another in the belief that such competition produced more innovative schemes, provided a catalyst for the development of ambitious plans for economic regeneration which, if not funded immediately, would help local authorities to decide priorities and implement new working practices, and assisted in securing the interest and involvement of the private sector. Such initiatives were also part of a broader competitive philosophy employed by the Conservatives which challenged the role of local authorities in the policy process and which redefined the division of power between the public and private sectors (see Oatley and Lambert, 1995; Painter, 1998). As Oatley writes:

> The introduction of City Challenge (May 1991) . . . and subsequent initiatives based on competitive bidding was aimed directly at changing the practices of local authorities and other agencies in the locality involving a shift away from urban managerialism towards new more entrepreneurial forms of governance and a reorientation of policy towards encouraging competitive business and competitive localities. (1998: 30)

The adoption of the competitive principle by the RDC in the construction of the RC initiative appeared to be based on a number of similar beliefs about the 'benefits' of direct competition. Running through the early documentation and publicity for the initiative is the assertion that competition would reward the 'best' projects and that it would lead to more innovative and entrepreneurial approaches to rural regeneration. According to the RDC:

> RC has encouraged a holistic, entrepreneurial approach to rural regeneration . . . the competitive nature of RC, including the local and national stages . . . generated a high quality of bids.
>
> . . . the benefits of RC could not be achieved without its competitive element. (RDC, 1998: 8–10)

This was further emphasised by RDC RC staff:

> The perceived benefits [of RC], then, were that competition would sharpen up people's approach, the process would enhance the quality of presentation . . . and bring forward large scale, more visionary schemes. (RC manager, 1995)[2]

---

2   All the quotes from Rural Challenge and RDC personnel, not otherwise referenced, were from interviews conducted as part of an ESRC funded research project (award number R000221853).

Again, in defending this approach to funding rural regeneration, the claim amongst those responsible for RC within the RDC was that the potential benefits of preparing a bid for RC outweighed the waste of effort that 'losing' would involve. Indeed the belief was that projects not selected for RC funding would not be 'wasted' but would be in a good position to seek out and qualify for alternative funding. Moreover, the process in itself would improve the skills and sharpen the priorities of local authorities, making them more likely to win awards in the future. In a review of RC schemes the RDC write:

> Some partnerships are considering the future after RC funding ceases even at the bidding stage. For several schemes, RC has provided the launch pad for securing additional public and private funding for a wide range of other projects. (RDC, 1998: 9)

It is not the intention here to provide a lengthy discussion of the implications of competition funding for the regeneration of rural areas – evidence from the local level contradicted many of the assumptions regarding the benefits of a competitive approach and considerable concern was expressed even amongst RC winners about the relationship between competitive bidding and need (see Little and Jones, forthcoming). What is important here is to note the introduction of competition into rural funding as part of a shift in rural policy-making and one which is significant in the new governance context. RC was recognised as different from previous forms of regeneration funding (at least in the early stages of its introduction) because of its use of staged competition. Those applying for awards, moreover, saw competition as responsible for a number of changes in working practices within the public sector. Some RC project managers spoke, for example, of competition increasing the pressures on individuals in terms of both meeting deadlines and producing well-prepared projects. The presentations required for such competitions were seen as particularly stressful and seemed to some to 'get in the way' of the real aims of rural regeneration. Such methods were, it was claimed, likely to reward flashy presentations over sound, but possibly conventional projects. As one project officer said:

> But I think the problem with bidding is, is it where the absolute need is? Then you stand back from it . . . I wouldn't like to differentiate between different places. Is [this] the area of most need? Probably not, we just happened to put the best bid together. (RC project officer)

In some cases RC project officers described how their authorities were 'gearing up' to challenges in a major way and while this style of funding was not necessarily supported by those working in the public sector, there was a general feeling that it was here to stay. Some RC project officers noted that while the RC was an initiative conceived under a Tory government and one that adhered strongly, as argued above, to Conservative ideology, it was retained by the incoming Labour administration of 1997. Its subsequent demise did not, it was felt, ultimately challenge competition as a means of allocating funding.

Among the other facets of new governance embodied in the RC initiative is the emphasis on economic priorities within the regeneration process. While the RC documentation stresses that the initiative aims to secure/foster the economic *and* social regeneration of rural communities, the underlying message is that it is through the development of the rural economy that lasting regeneration will be achieved.

> Partnerships . . . were encouraged to propose schemes which integrated business development, training, property and social projects. A number of schemes focused on environmental, social and community development activity *in order to support the economic regeneration of an area.*
> (RDC, 1998: 8; emphasis added)

The bidding documentation and the funding decisions made by the RDC appear to promote the view that social aims will follow on from successful economic regeneration and that it is the latter that should receive the greatest attention within the overall proposals. Certainly those responsible for devising and presenting RC initiatives at the local level were of the view that 'hard' economic development schemes were more likely to win support than 'soft' social projects. The belief circulating during the early stages of the RC initiative was that proposals containing major capital development – preferably involving significant job creation – stood the greatest chance of being selected. The view amongst competitors was that 'Ministers like something they can stand in front of and cut a ribbon to open'. Many talked of tailoring their projects to reflect this prioritisation, even though this may not fit with the experience of need within their local area.

Related to this emphasis on economic need within RC is the encouragement given to the private sector. This encouragement, already discussed as indicative of the changing nature of governance, is apparent in the 'rules' of RC and, less formally, in the practices of those bidding for funding. The presence of a major private-sector player in a bid is acknowledged to improve significantly the chance of success. Ironically this has remained the case through the course of the RC initiative despite two important and widespread observations. These are, first, that the size of the private sector in rural areas often precludes their involvement – in rural areas there often aren't hundreds of private-sector agencies and businesses queuing up to buy into local economic development and where they do exist, they often do not have spare resources to commit to speculative initiatives such as RC. Secondly, evidence during the initial rounds of RC showed that the private sector could be a difficult and unreliable member of any partnership. This again was reinforced by the size of the private sector in rural areas (individually and collectively) and frequently resulted in partnerships falling apart or having to be reconstituted mid-project.

Another key development in the shift from government to governance noted earlier is the increasing involvement of a broad range of non-elected organisations in the decision-making process. Increasingly policy is seen to be formulated and implemented by partnerships of agencies and by the coming together of the public and private sectors. Again this trend is enshrined in the RC initiative. Bidding for RC awards can only be undertaken by *partnerships* of agencies. Indeed, the RDC even specifies the composition (albeit in general terms) of such partnerships: membership must include representatives of the public and private sectors and the community. Again the assumption generated (though not necessarily justified) from urban regeneration that partnerships are 'undeniably a good thing' (Peck and Tickell, 1994: 251), has not necessarily been supported by the evidence from rural areas. RC project officers spoke of the difficulties of securing lasting partnership arrangements and the problems associated with partnerships that fell apart or existed in a state of permanent flux. In Middleham, North Yorkshire, one of the first-round RC winners, for example, the original partnership board changed so significantly that three years into the project only two of the original 16 members of the partnership board remained the same. Project representatives also talked of the construction of false or paper partnerships for the purpose of bid preparation and of the difficulties subsequently caused for the implementation of schemes and the division of responsibilities. Nevertheless, the embracing of the partnership principle is seen as a further sign of the shift in the nature of local state activity in rural areas and of changes in the policy process.

The types of changes in the governance of rural areas described here but also observed elsewhere may have been seen to have problematic implications for rural areas and their adoption may have been contentious and contested at the local level. Critical reports of such changes have been published elsewhere (see Jones and Little, 2000; Little and Jones, forthcoming). The point here is that, contested or not, some of the sorts of changes equated elsewhere with a shift from government to governance are being observed within the rural policy process. It is clear in the design of the RC initiative that policy-makers are being influenced by the same central state requirements – financial, political and ideological – as have driven urban regeneration policy. As argued, these may take different forms locally and may result in differing outcomes but are linked to a common direction. Whether the local celebration of difference can sustain real challenge to the broader principles of governance is a question that is beyond the scope of this chapter but one which, nevertheless, is important to the understanding of the future impact of rural policy-making.

Gender and Rural Challenge

This next section examines the relationship between gender and new rural governance in the context of the key themes raised above. It looks primarily at the implications for the gendered experience of and involvement in rural policy-making of the kind of changes associated with new forms of rural governance. The main changes as identified above can be distilled into three main areas for this analysis. First, the prioritisation of economic goals; second, the increased involvement of business and the private sector; and third, the emphasis on competition and new styles of working, including the use of partnerships for the delivery of rural regeneration.

The prioritisation of economic goals

The discussion above has argued that the changes taking place within the policy process have, as exemplified in the case of RC, encouraged the prioritisation of economic objectives within local regeneration. This has, there is no doubt, been tied to the increased involvement of the private-sector business community as discussed below. It is also manifest in the decisions around the *types* of projects seen as appropriate for achieving rural regeneration. Table 6.1 includes a list of projects funded under the RC initiative and demonstrates the dominance of large developments with an economic function as the focal point of most schemes. In favouring major capital projects involving large 'bricks and mortar' schemes, as

Table 6.1 Rural Challenge projects funded 1994–6

Location of project	Project proposal	Total value (£ million)
Boughton Pumping Station, Nottinghamshire	Restoration and revitalisation of Boughton Pumping Station. To create within it craft workshops, offices and other workspace options, together with exhibition, community and recreational facilities.	2.5
Middleham, North Yorkshire	Development of the town's main industry of racehorse training to provide employment and training and to encourage tourism.	2.7
Bishop's Castle, Shropshire	Conversion of factory into business starter units and expansion of existing businesses. Provision of capital grants for shops. A package of training and educational resource provision. Service provision and childcare.	5.5
Brookenby, Lincolnshire	Redevelopment and marketing of a technical park, improvement of infrastructure, roads and street lighting. Provision of a family resource centre, youth club and recreational facilities. Vocational training and adult education.	2.5
Miora, Leicestershire	Provision of craft workshops, retail units and commercial/office development. Establishment of a National Forest training and visitor centre.	3.0
Watchet, Somerset	The development of a marina facility. The establishment of the harbour, esplanade and wharf as the centre of activity and focal point of the town.	3.9
Stainforth, South Yorkshire	Town-centre redevelopment including new work spaces. A package of training and education facilities. Sports and recreation provision.	3.6
Jaywick, Essex	Infrastructural improvements including transport. Provision of workshops and training. Upgrading of recreation and play space. Development of crime prevention schemes.	2.8
East Sussex	A programme of measures to regenerate the woodlands of East Sussex RDA and provide information and training.	3.2
Swaffham, Norfolk	Development of a business park and Eco Tech centre. Provision of training related to the new businesses.	13.0
Cornwall and Isles of Scilly	Creation of 40 'Signpost' points at village locations to provide information for local people and tourists.	2.8

Table 6.1 *cont'd*

Location of project	Project proposal	Total value (£ million)
Bakewell, Derbyshire	Development of new livestock market and rural enterprise centre. Redevelopment of old market. Provision of community centre and training programmes.	5.2
Great Torrington, Devon	Building of a new multimedia centre, tourist attraction and community facilities linked by shuttle bus.	1.9
Rural Suffolk	Housing for single unemployed people and transport to training facilities.	2.2
Rochdale Canal	Canal technology centre and market area at Hebden Bridge.	2.9
Somerset Rural Youth Project	Provision of mobile information, leisure and training facilities.	1.5
Saltburn by the Sea	Refurbishing town centre including redevelopment of an old building to provide studio workspace and facilities for tourists.	1.8
Ashby Canal	Restoration of canal, development of residential and business premises, establishment of wildlife corridor.	4.3

Source: Little and Jones (2000)

noted above, RC was, it may be argued, promoting a very masculine interpretation of development. While qualification for RC funding necessitated some sort of community initiative, this was frequently a secondary part of the overall project – an 'add on' and seemingly dispensable element of the scheme. As one of the RDC managers put it:

> It's much more difficult to show what you've achieved in terms of a counselling project or a music project. But if you can say 'I've built this' people can see it and see where the money has gone.

The prioritisation of economic goals set the tone of the local RC projects from the start, not only in terms of the design and funding of schemes but also in relation to the personnel involved and the requirement for particular skills and expertise. It was also the case that the immediate employment created by RC awards, to actually build these major regeneration schemes, was frequently connected to male-dominated professions – surveyors, builders, engineers etc. In relation to the long-term job opportunities provided by the RC projects, although the majority included service-sector, often tourism-related, employment, this was generally dependent on the redevelopment or reinstatement of former industrial uses, again with very male-dominated forms of employment.

Some of the RC project officers were uneasy about the RDC's prioritisation of economic elements of their schemes. One – an initiative rather out of character with others that were funded – was designed to provide access to education, training and leisure for young people and was run by people with a background in 'youth service' rather than regeneration and economic development. A representative from this project talked of the tensions felt in reconciling the need for an 'entrepreneurial business element' with a 'youth service philosophy'. He saw the flagship culture as 'aggressive and masculine' and often at odds with the sorts of ethos and ethics behind the RC initiative. The fact that this project had actually been funded was quite a surprise to the representative – his view was that it had only won an award as they had pushed certain elements of the project, notably the purchasing of a large amount of 'high-tech' equipment and a number of vehicles which gave the project a more high-profile image.

Although it is important not to fall into rather essentialist associations of economic development schemes as 'masculine' and community or social schemes as 'feminine', it was clear from the analysis of RC projects that the elements that were directed towards welfare or community development tended to involve more women – in terms of both the organisation and the objectives of the initiatives. As one RC partnership member put it:

> At the higher level, the partnership board, the project deliverers, are all men – well, all the chief exec types representing organisations are all men, apart from one who represents a community group. Women tend to be involved with the community or education.

Frequently it was suggested by those working on RC projects that women were more 'suited' to the community parts of the schemes – they had the skills needed to organise and enthuse village residents and understood how the community worked in a way that men did not. Such skills were, however, seen as less appropriate to the generation of business ideas and the creation of 'flagship' economic development schemes. So –

> On the board it's all men at the moment. Although we have some enthusiastic women in the village . . . on the board it's all men.

Significantly, as noted above, in the implementation of RC projects it was often the 'community' element that was postponed, downsized or even written out of the final scheme.

One of the early RC projects was a bid by the village of Middlёham for funding to expand racehorse training facilities, the village being the centre of a highly successful local racehorse training business (see Middleham Key Partnership, 1994). The bid included funding for the provision of new all-weather horse training and veterinary facilities, educational training for stable hands and other local people together with environmental improvements in the village and a new community centre. Interestingly, while the horse training facilities were instigated soon after the partnership had secured RC funding, the community centre ran into a series of problems and there was talk of abandoning it completely. This would have been highly disappointing to the community as a whole for while local trainers and other businesses benefited directly from the injection of money into the training facilities, the main benefits of RC to the community were due to come from the new community centre. Finally, after lengthy negotiations, the community centre did get the go-ahead, but its completion was much delayed and followed long behind other parts of the project.

The increased involvement of business and the private sector

Linked to the proliferation of major capital/economic development projects within RC is the growing involvement of the private sector in rural regeneration. This, as already argued, is a feature of the shift to new forms of governance and represents not only the broadening of participation in policy-making but also the domination of economic policy and business interests within the local state. Here again there are important and far-reaching implications for gender relations within both the process and outcomes of policy-making. The evidence from RC was that, as Tickell and Peck claimed in their Manchester work, women tended to be more poorly represented within the private sector and within the various structures and organisations, formal and informal, through which 'business interests' operated. The small size of the private sector in rural areas appeared to reinforce the status of 'key actors', with leading business people – generally men – assuming a number of influential roles locally. Thus in the RC projects investigated there was evidence of collections of powerful individuals associated with the partnership boards. These were people who were well known in the area and often involved in several local ventures. Partnerships were frequently chaired by such individuals, the vast majority of whom were male. As one RC chair described 'his' board:

> Well, currently I'm the chairman. Slightly forced on me really. Right at the beginning I was involved with Rob, Mike and Tony – the ones in the original town council. They were the ones who put the idea together. And then it was decided to bring in more local business people to try and see it through. That's when I was approached – mainly because of my involvement with other things; with Town Twinning, with other associations like Round Table and Rotary which I was fairly active in. It's the old adage – ask a busy person . . .

The stipulation that RC schemes had to be devised and put forward by partnerships further reinforced the role of these male-dominated networks and informal organisations. In searching for the requisite number of partners, the organisers of RC were frequently dependent on local key actors and on the associations that were supported by them. Contacts within the business community were seen as an important qualification of partnership membership. Such contacts, as already noted here and by other authors (Watson, 1990), are generally more easily established by men.

Even at more public meetings in connection, for example, with ascertaining community responses to proposals within the design of RC schemes, the influence of a masculine business culture was seen to dominate. As one (woman) member of a partnership board explained:

> It's all very cultural – 'do you want to be on Tim, do you want to be on Dave?' so I put up my hand and [say] 'excuse me don't you think you should have a woman rep?' And it's pushing a point all the time – that there should be women reps on the committees. But they do, women get overlooked.

Competition and new styles of working

The argument that with RC major shifts were taking place in the styles of working amongst policy-makers was supported through the case-study evidence associated with the project. This is particularly true in relation to the use of staged competition as a means of allocating funding but was also apparent in other ways in the working practices of those involved in the initiative at the local level. That these new practices were gendered was also clear. There is no reason why competitive bidding in itself should marginalise women's involvement in the decision-making process. However, the assumptions made regarding the ways decisions were arrived at – especially the belief that the funders favoured more entrepreneurial and 'hard hitting' presentations – encouraged, so it was argued, a very aggressive and masculine approach to the bidding process. The RDC's view that competition would 'sharpen up people's approach [leading to] larger, more visionary schemes' was reflected in the local response to the presentation of their schemes.

A 'successful' RC scheme needed to win at both local and national level. The national competition, in particular, was seen as a high-profile event in which teams presented their schemes at the offices of the Department of the Environment (now Department of Environment, Transport and the Regions) before commissioners of the RDC and a Member of Parliament. Teams prepared for these events thoroughly, anticipating aggressive questioning and tough opposition. One of the presentation team from a winning project talked of their experience in the following terms:

> Oh we were slick, don't worry. Four of us went on a presentation course, 'specially. You have to have the technical side and you have to play the media. You have to be totally on top of your project so that when you are asked questions you've got all the answers. And if you haven't got the bloody answers you have to be plausible.

While another remarked:

> We certainly prepared ourselves thoroughly. Working out questions we'd be asked and what would be the best answers. It went like clockwork. They in their grey suits, sat one side of the table and we, in ours, on the other . . . The DOE offices are a soulless 1970s building. The long room, the long table, the grey suits seemed a long way from Bakewell [the project base].

Although many professed to concerns regarding the appropriateness of using competition to allocate funding for rural regeneration in this way, the involvement in the presentation and the achievement of winning coloured the views of some of the participants. In describing their preparation and final triumph project officers frequently used military and sporting analogies – 'fighting off the enemy', 'thrashing the opponents', 'fielding their strongest team'. Where one county had been successful in two successive rounds of the competition they reported their 'victory' as 'two-nil to Somerset'. Losing teams were, unsurprisingly, more critical of the competition process. A woman from a 'failed' bid described it as 'macho' and 'alienating'. She stressed, in particular, how the project team (from a community council and mostly women) felt the competition (and the assumptions made about how it should be managed) to be a very disempowering experience in which the original ideas were changed out of all recognition.

The more confrontational approach to rural regeneration policy was also reflected in the preparation of RC bids away from the actual presentational aspects. Again, many of the observations concern practices which in themselves are not necessarily exclusive to men but which nevertheless were seen as aggressive and masculine. Some, moreover, had practical implications which favoured ways of working generally associated with men. For example, in many cases the preparation of RC bids, particularly during the final stages, had been frantic and time-consuming. Those involved had often had to work long hours in order to meet the tight deadlines imposed by the RDC. Such flexibility in working hours is often not an option for women who are usually the primary childcarers and domestic workers within the household. Constructing a bid, as noted earlier, involved seeking out possible partners, convincing people that they wanted to be involved and generally 'selling' the ideas. As already mentioned, this required a knowledge of local networks and business organisations. It also relied heavily on an aggressive strategy in which possible partners could be persuaded to get involved and convinced of the benefits of the project. Again, this kind of approach can be equated with more masculine ways of working.

Perhaps less controversial is the argument that the construction of RC projects and, in particular, the emphasis on major capital schemes involving large-scale building and engineering work, reinforced the masculine nature of the policy and ensured that the majority of project officers, appointed to oversee RC projects, were men. Of the 23 projects funded over the four rounds of RC, only three project officers were women. These project officers varied in terms of background and work experience but there was a dominance of economic developers and engineers. One RC project officer summed up what he saw as the skills for the job:

> An effective communicator and public consultee; sort of motivator; a hard-nosed business developer and letting agent; a business manager; a clerk of works and a site manager; an architectural specifier.

While another related this to his own position:

> My background is that I'm a chartered secretary and a chartered builder and other things as well and that stands me in good stead for servicing the board meetings and everything.

As projects became broken down into various parts, the association between 'hard' economic schemes as male and the 'softer' community initiatives as female was clear and the assumption that women's skills, experience and preferences were best suited to the more social/welfare part of the project as opposed to the building/development work very apparent. Before turning to focus on this in more detail it is important to make the point that the male bias of RC projects went beyond the local scale. At the time when the RC research was undertaken the majority of RDC officers were male – especially at the senior levels. Project officers interviewed for the RC research described the RDC as a highly traditional organisation with a stereotypical attitude towards gender and to women's role in the policy process.

Gender, politics and the rural community

Throughout this chapter the association, both empirical and conceptual, between women and community in the formulation and implementation of rural policy has been stressed. The representation of women on parish councils has been discussed as an indication of their more significant role in community-based structures of decision-making, while the examination of RC has documented the persistence of assumptions that women's interests and skills are most suited to the community sphere of policy. Clearly, the gendered nature of involvement must be taken into account in the analysis of rural policy at the local level. Such analysis must also incorporate an understanding of the role of the *informal* process of policy- and decision-making. So far the focus of this chapter has been very firmly on the formal mechanisms and practices of rural policy. Calls for more attention to be paid to the notion of the 'active citizen' and to the role of the community in policy, together with the long history of voluntary activity within rural communities (see Chapter 4), requires that we also think about the influence of those operating outside the formal structures in the formulation and implementation of rural policy and, specifically, the ways in which such involvement is also gendered.

As noted above, women are clearly increasing their representation in community politics through involvement on parish councils; the extent to which this is affecting the wider gender relations in local policy is, however, less clear. Past studies of power in rural areas have noted the importance of key 'positions' or elites within the village – these are often not political positions *per se* but they can carry considerable weight within the policy process. Such studies have also indicated that these elites are frequently men: local landowners and farmers, for example, and local vicars. Although the gender balance within such occupations is changing, they are still heavily dominated by men (see Woods, 1997). Seymour and Short's (1994) research on the rural clergy, for example, has shown that while women are heavily involved in the rural church, the official roles tend to be held by men. This is particularly true of salaried positions in the rural church; Seymour and Short found that only 5–6 per cent of paid staff were women as opposed to 25 per cent of the unpaid staff.

> An easily recognisable, professional representative of the church was over 13 times more likely to be a man than a woman. (Seymour and Short, 1994: 47)

These elites may not hold formal offices but can still be influential in setting the political agenda in rural areas – they have been shown to exert a significant power over issues such as local planning decisions. They may also act as old-style benevolent 'lord of the manor' figures, providing funding for local 'good causes' and volunteering for certain organisational roles within the village.

The most effective medium through which women influence the political life of the village is informal, community-based groups. Such groups, set up, for example, to provide childcare or support for an elderly person or to raise funding for a village facility, are rarely political in a formal sense. They do not usually have a campaigning role and are frequently about 'making do' in the absence of some officially provided service. Often groups established to meet a short-term and localised need become much bigger and more prominent than the initial intention. Some may even be adopted by the state, receive funding and become an element of formal service supply. Participation in such groups is common amongst village women, as Chapter 4 has discussed, and reflects strong and entrenched views about both the nature of community work and the role of women.

Interestingly, this involvement by (mainly) women in community groups in rural areas is not conceptualised, from either within or without, as political. In some ways the setting up of a childcare facility could be compared to the overtly political acts of urban women in the 1980s – a form of community-based activity which is about responding to local needs but at the same time drawing attention to the inadequacies of formal provision. The ethos of 'self-help' so central to the construction of rural communities and rural people, is drawn on by both residents and policy-makers alike to depoliticise the establishment of, and involvement in, such groups. Acting independently, looking after others and ensuring the well-being of other members of the community constitute a powerful aspect of the dominant rural ideology and one that is often employed in descriptions of the rural community and its superiority to its urban counterpart. Thus participating in self-help is about protecting the village community, its sense of uniqueness and its separateness from the urban. It is seen as a 'natural' part of living in a village, not a political gesture prompted by the deterioration of essential services.

It may also be argued that its construction as 'women's activity' also serves to depoliticise voluntary work. While we must be wary of creating a tautologous argument – that voluntary work is not political because it is done by women but at the same time women get involved in voluntary work because it is not seen as political – it is clear that the gendered and non-political natures of self-help do reinforce one another. While contributing at times enormously to the well-being of individuals in the rural community, there is a sense in which women's voluntary work in the village is an extension of their domestic tasks and as such it is unrecognised and unrewarded (financially, at least). It is thus conceptualised, by both the women who participate and the wider community, as belonging more in the private sphere than the public.

The true political nature of voluntary work derives largely from its role in replacing state-provided services. For example, the high numbers of village women 'helping out' in the local primary school identified in my Avon-based research have already been referred to. The willingness of women to act as unpaid staff was clearly crucial to the activities of the school and even to its continuation – a fact that was not lost on the parents (mothers) themselves. Voluntary activity was saving the local education authority money, allowing them to keep the school running and, furthermore, not having to pay for children to be transported to another school. This situation may not have been engineered cynically by the policy-makers but the wider implications of the voluntary work were clearly political, even if the motives for action on behalf of the volunteers were overwhelmingly personal.

Involvement in local voluntary activity can also be political in terms of its ability to empower individuals and groups. While, as noted above, much voluntary activity is inspired by motives that are not perceived of as political, a by-product of the action itself can be the politicisation of participants and/or a raising of their consciousness in relation to broader social needs and opportunities. Frequently cited in relation to women's involvement in formal politics via the route of voluntary work are examples such as the miners' wives' support groups from the 1984 miners' strike in Britain and the women's anti-nuclear peace camps of the early to mid-1980s. Many other, less well-known but equally empowering examples exist in rural areas where women have discovered a skill for organisation and activism and a wish to put such skills to further use in more formal structures of power.

The question that needs addressing here, however, is whether these examples of community action by women constitute (or are likely to constitute) a threat to the established channels of power in rural areas or to the mechanisms and practices through which power is mobilised. It has been argued that while voluntary activity or self-help *per se* may not be constructed as 'political', it does have, in many cases, a political context and message. In addition, participating in such activity can help to politicise individuals in the rural community. It also has to be argued, however, that self-help is long established in rural communities and has consistently made a vital contribution to service provision and social welfare in such communities. Women have been as important in the past to this informal activity as they are in the present day. Yet the gendered nature of rural politics and policy-making has changed relatively little. Of course as a part of a wider society, village communities have reflected something of the changing political role of women over the past century and cannot be said to be immune from the shifting nature of gender roles and gender relations in policy-making generally. It is difficult to argue, however, that women's greater involvement in voluntary activity within rural areas has given them any real additional power or enabled them to exert a special influence on the policy process. Thus despite women's valuable (often essential) contribution to policy at the village level, the gendered nature of rural decision-making as outlined above continues with little change. Indeed, for the most part, the way in which voluntary work is constructed and undertaken only serves to reinforce the power balance between men and women in rural communities and to protect established patriarchal gender relations.

Conclusion

This chapter has articulated a clear need for a feminist analysis of power and policy-making in rural areas. It has identified the serious neglect of gender in terms of the understanding of formal structures of rural decision-making and the broader operation of political power and governance at the local level. The absence of a gender perspective on rural political power has both stemmed from and contributed to the resilience of a belief that 'gender issues' in relation to governance and the policy process are essentially urban.

The chapter has gone on to demonstrate that despite the absence of research in the past, the emerging interest in new rural governance has provided an important opportunity to begin to interrogate the gendered nature of the distribution and outcome of political power in rural areas. This has involved a detailed examination of rural governance including not only the personnel involved but also the practices, mechanisms and cultures of policy. The chapter has paid particular attention to the role of economic development in the context of local rural policy, arguing that the emphasis on business and economic goals has had profound implications for the direction, content and, importantly, the gendering of rural power and governance.

There are some signs, as in capitalist economies generally, of an increase in women's representation in political structures and positions within rural areas; a sign, perhaps, of a shift in the gendered nature of rural power. At the time of writing, a popular rural publication in the UK (*Country Illustrated*) had announced the elevation of three 'leading ladies' to the top offices of high-profile countryside agencies (the National Trust, the Co-operative Wholesale Association and the Environment Agency). While the publication claims that these appointments signal a breakthrough for women as far as rural policy-making in the UK is concerned, it should be noted that these agencies are not located in rural areas. Neither is it clear whether the three women concerned would see themselves as 'rural women'. The extent to which they can be seen as role models for other women to become involved in rural policy-making must be doubtful. The article itself recognises the continuing absence of women from more traditional countryside organisations such as the Game Conservancy and Forestry Commission (Barber, 2000). What is clear is that the involvement of women in the politics and policy-making of the countryside has not yet begun to challenge the construction of rural women as primarily wives and mothers. Decision-making in the countryside may increasingly involve women at a range of different levels but it is rarely directly about either gender or women's issues. Maybe that stage of political activism is yet to come, but there are few signs of its imminent arrival.

7 | Gender, sexuality and rurality

'If we were city women,' Alice said slowly, 'we'd have a completely different life. It's being country women that makes it so difficult. Even if I moved to the city I'd still be a country woman now. I'd still feel visible.' (Trollop, 1989: 220)

Introduction

Apart from some reference to 'queer theory' in the discussion of feminist perspectives in Chapter 2, discussion of sexuality in this book has so far been restricted to heterosexuality. This is for good reason, reflecting the dominance of conventional notions of sexuality within rural lifestyles and cultures. As has been stressed in previous chapters, dominant constructions of rurality reinforce heterosexual relationships to the exclusion of other forms of social and sexual relationships. The strength of the 'family' in rural society and community both stems from and contributes to the dominance of heterosexuality and while the privileging of male–female over same-sex relationships is clearly a feature of social relations more generally, the normalisation of heterosexuality is, it is argued, particularly powerful in the rural context (see Valentine, 1997).

The strong dominance of heterosexual relationships in rural society does not, however, mean that the discussion of sexuality is a non-issue. It is important to look behind the assumption of heterosexuality and to explore in more depth the powerful association between conventional forms of sexual relationships and rurality. To explain this association simply by identifying the importance of the family is clearly inadequate, and the particular role of rurality in dominant sexual beliefs and practices must be unpacked. Moreover, while same-sex relationships may be largely absent as part of lay and popular constructions of the rural, gays and lesbians do live in rural areas as couples and on their own. Their various experiences of rural life need to be explored, particularly as they relate to ideas of otherness and social exclusion within rural society. The lives of gays and lesbians living in rural communities can help us to understand more about the recognition and experience of marginality as felt by a specific identity and also about the workings of rural society generally in its support of heterosexual gender relations.

Geographers interested in the relationship between sexual identity and space have worked, as noted in Chapter 2, mainly in the urban context (for a review of this work see Binnie and Valentine, 1999). They have discussed the ways in which certain spaces can become associated with gay and lesbian identities – obvious examples being the Castro area of San Francisco (see Castells, 1983; Lauria and Knopp, 1985) and Soho in London (Binnie, 1995). This work has also stressed the importance of conceptualising sexuality as a process, formed in relationships, 'rather than as a stable, immutable thing' (Mitchell, 2000: 175). Part of understanding the relationship between sexuality and space involves recognising the differing ways in which people perform their sexual identities in different places. Looking at the rural, then, requires that we think about the varying ways in which gay and lesbian identities can be constructed and lived and how the varying socio-cultural characteristics of places impose differing constraints on the *process* of sexuality.

The examination of homosexual lives and identities also requires that we take seriously the importance of (and debates around) embodiment. As discussed in Chapter 2, the body has become a key focus in the performance of identities and, as such, in our academic interest in performance. The performance of sexuality has, in particular, shifted attention to the body as a site of power and resistance. The body as it is used in the reinforcement and disruption of different forms of sexuality is thus a central consideration in the examination of the construction and marginalisation of rural homosexual and heterosexual identities, as will be seen in this chapter. Before looking specifically at the embodied performance of homo- and heterosexuality in rural communities, the chapter starts with an exploration of the question of representation in terms of the relationship between sexuality and rurality.

Rurality and representations of sexuality

The assertion of the dominance of heterosexual identities in rural communities has tended to mask the existence of a range of homosexual imaginings of the countryside. But while gay and lesbian relationships may be marginalised by dominant constructions of rurality, the rural has been clearly evident in what Bell (2000a) describes as a symbolic resource within homosexual cultures. Bell stresses the importance of exploring not only the representation of homosexuality within dominant constructions of the countryside but also the different ways in which some gays and lesbians have made use of the countryside in the construction of particular forms of social and cultural identity. The intention here is to consider both of these aspects of the representation of rural sexuality. Separating 'heterosexual' and 'homosexual' representations of rurality may be an overly simplistic way of identifying different sexualities with rural constructions. It is not intended to imply that there are two, and only two, sharply divided rural sexual imaginings; as will be explained, these major 'categories' mask more complex and shifting representations of sexuality. It does, however, help in making the main·point of this section, namely the dominance of heterosexuality and the exclusion of homosexuality in contemporary rural cultures.

Heterosexual representations

The wealth of literature that now exists on the popular representations of rurality in Britain and other western countries stresses the power afforded to certain groups and individuals as a result of their (conscious and unconscious) ability to 'fit in' to mainstream visions of rural life (this literature has been discussed at some length in earlier chapters). This power reinforces conventional hierarchies in a self-sustaining process which serves to marginalise certain identities and ensures the exclusion of those groups and individuals who remain outside or 'other' to the dominant constructions. As a result rural society, it may be argued, becomes less and less diverse or tolerant of difference. The much-applauded sense of community believed to operate in rural areas only supports a narrow set of values and rural imaginings, providing a smoke-screen for a deep-seated intolerance of views and people who do not conform to the 'country ways'. Within this context homosexuality is clearly 'other' to mainstream representations of rurality. Gay and lesbian identities are absent from all aspects of rural life, having no place in the very consciously heterosexual spaces of the rural community.

As noted above, space and sexuality are inextricably linked. By looking at the representation of various social spaces in the rural community it is possible to recognise the clear assumption of heterosexuality as the dominant (and possibly exclusive) form of sexual identity. One of the spaces central to conventional images and imaginings of the rural, in Britain in particular, is the church. Although the role of conventional religious worship is declining in contemporary society, the church remains physically and symbolically a powerful element of rural life. Notwithstanding its changing social and political significance in the operation of rural community life, the church still features highly in conventional depictions of the village as a key part of the rural idyll. Nowhere, it may be argued, is the assumption of heterosexuality so unwavering as in the rural church. While some attempt has been made to break down the image of the church in general as patriarchal and highly traditional, the rural church remains resistant to change, particularly in relation to ideas of gender equality (see Seymour and Short, 1994). The church as a family space is particularly strong in rural areas – there is still a tradition of families attending services at Christmas, Easter and harvest time together, for example, and the classic church wedding still centres on the *rural* church.

Many of the other key spaces of the village reflect, as noted in Chapter 4, highly conventional gender relations. Such spaces also reinforce the dominance of heterosexuality. Public spaces such as the village hall, for example, generally encourage family activities aimed, supposedly, at being inclusive but in essence suppressing difference and supporting 'the same' over 'the other'. Those 'gender-specific' events that do take place are highly traditional (for example the Women's Institute or Mothers' Union) and are not designed to challenge the primacy of the family. External spaces of the village such as the village green support activities and events that celebrate the unchanging nature of the village and the harmony of social relationships. They provide space for the expression of family cohesion and are fiercely protected in the face of transgressive behaviour which might unsettle accepted and acceptable forms of social interaction. Important village festivals and customs frequently revolve around (or at least include) some sort of fertility rite, emphasising not only traditional gender relations but also heterosexuality.

The celebration of May Day, the crowning of the May Queen (the heterosexual version!) and the celebration of spring are obvious examples. A specifically Cornish (and very ancient) version of this spring fertility festival is found in the Helston Furry Dance – a dance involving local couples in a celebration of spring. The Furry Dance is described by MacFayden and Hole:

> At 10am the children dance, all in white and wearing the traditional lilies . . . Then, exactly at noon, the principal dance begins. The Mayor goes first, wearing his chain of office, and behind him come men and women in couples, the men in their morning coats and the women in their prettiest summer frocks. They dance all through the main streets and into gardens, shops and houses . . . to bring the luck of summer to the owners and tenants and drive the darkness out of winter. (1983:42)

Conventional constructions of masculinity and femininity and of sexuality can also be observed in other 'local' village festivals. In Devon, for example, there is an annual barrel-rolling festival in celebration of 'Guy Fawkes Day' in which huge lighted barrels of tar are carried round the streets of Ottery St Mary by local men. The idea is for men to run, in teams, as far as possible carrying the burning barrels, and every year men (both taking part and those getting involved as spectators) require medical treatment for burns and other injuries. The event rests on highly conventional displays of masculinity including fitness, strength and fearlessness.

Agg and Phillips (1998) discuss the traditional gender relations portrayed in stereotypical images of rural society. They draw attention to the conventional representations of masculinity and femininity evident in, for example, advertisements for 'country clothing' and other products associated with rural life. These conventional gender images, I would suggest, are heavily inscribed with an assumption of heterosexuality. Men are rugged and weather-beaten but with a clean-cut appearance, while women are coy, demure and rather wholesome. Men and women frequently appear together as couples, or with children. These images do nothing to disrupt or challenge conventional representations of heterosexuality and appear to ground such images strongly within their rural context. A similar observation can be made in relation to images in specific 'rural' publications, the obvious example being *Country Life*. Again, while Agg and Phillips (1998) use these images to make a point about gender relations, they can also be seen as excellent examples of the reinforcement of the assumed (and actual) link between rurality and heterosexuality. The women featured in particular in the 'girl of the week' section are 'feminine' and chaste (see Figure 7.1). They are generally 'advertised' with a long pedigree including a smattering of upper middle-class and aristocratic relations, reinforcing the importance of family connections and the assumptions that the family line will be continued.

Bell and Valentine (1995b) note the presence of a strong moral code in constructions of rurality and suggest that the extreme dominance of heterosexuality in rural areas is an important part of this morality. They argue that the portrayal of rural life as simple sustains a moral economy of the countryside which conflates with hegemonic sexualities. Radio and television shows as well as country and western music reinforce simplistic depictions of sexuality in which

> the only potentially counterhegemonic displays of sexuality are of the cheeky 'love-amongst-the-haystacks' variety, easily absorbed and rarely mentioned.

COUNTRY LIFE

Vol CXCV No 11, MARCH 15, 2001

MISS ROSAMUND BENNETT

Miss Rosamund Bennett, younger daughter of Mr and Mrs Nigel Bennett, is to be married to the Honourable John Maitland, the Master of Maitland, son of the Master of Lauderdale, Viscount Maitland, and Viscountess Maitland, and grandson of the Earl of Lauderdale, who celebrates his 90th birthday on Saturday. Rosamund, aged 21, works for the management consultants Cap Gemini Ernst & Young. She is photographed here with her border terrier Rufus.

Derry Moore

Figure 7.1 *Country Life* continues to portray highly typical images of rural women. 15th March, 2001

They continue:

> In such an unreconstructed, essentialized environment, what place can there be for sexual dissidents. (Bell and Valentine, 1995b: 115)

This portrayal of rural sexual relationships as simple is clearly identifiable in the treatment of children/young people's early encounters with their own and others' sexuality. The classic image of the innocent rural childhood extends, it seems, to sexual relations and experimentation (see Jones, 1999). Laurie Lee's depiction of early sexual encounters and loss of virginity, for example, are imbued with a warmth and simplicity that are seen as closely linked – indeed part of – the idyllic nature of the countryside. Even a planned rape of a village girl becomes a harmless prank when contemplated by 'simple country lads' who are only innocently curious and at the mercy of nature's rhythms. The relationship between the countryside and sexual simplicity is also present in the more sinister issue of child abduction and abuse. Over recent years in Britain there have been a number of stories in the press about the abuse (and in some cases murder) of young children snatched from rural areas where they have been playing. The fact that such attacks have been perpetrated in rural areas has attracted attention and disbelief as children playing in the countryside have been portrayed as doubly innocent. Sexual deviance, especially that involving children, is seen as entirely alien to the rural community even though, as discussed below, the countryside has been associated with certain deviant (yet, crucially, harmless) sexual practices.

Just as the representation of rurality serves to reinforce a strong and conventional morality, so it also works to police moral standards. The tight-knit and 'close' nature of the rural community may help to provide support and care to certain groups and individuals but can also, as has been recognised elsewhere (see Halfacree, 1996), prove an intrusive and controlling force in the case of what are believed to be deviant practices or people. While in many rural communities counterhegemonic sexualities are, as the Bell and Valentine (1995b) quotation above maintains, 'rarely mentioned', there is an underlying acceptance (reinforced by the dominance of heterosexuality and family life, as discussed) of their inappropriateness.

It is not only the expression of homosexuality that is frequently frowned upon or outlawed in rural society but the existence of other sexual relationships that might threaten the harmony of the family. Affairs and other relationships deemed to be 'inappropriate' are also strongly policed by the rural community and are seen as threatening to the sense of stability within the countryside. The quotation at the beginning of the chapter is taken from Joanna Trollop's novel, A Village Affair (1989). It emphasises the fears of the lesbian lovers at the centre of Trollop's story concerning the visibility of their relationship within the village. The suggestion is, in the conclusion of the story and the eventual parting of the women, that the relationship might have succeeded/lasted in the anonymity of the city but that it did not stand a chance in the countryside. This is not only because the relationship was one involving two women but also because it was so 'public'.

Sexuality, if it has a place in the rural imagining, should be confined and conventional. While there are many examples of highly charged (and illicit) sexual relationships amongst literary and cinematic heroes, such portrayals of rural sexuality imply relationships generally doomed to failure and ultimately succeeded by more suitable, less passionate and more chaste sexual relationships. The triumph of *Far from the Madding Crowd*'s Gabriel Oak, or Colonel Brandon in *Sense and Sensibility* over their more exciting and dangerous rivals, says much about the inappropriateness of heightened displays of emotional and sexual relations in the otherwise tranquil countryside. Today the public performance of sexuality belongs in the city. The sexed body, homosexual and heterosexual, with its potential to disrupt and disturb is more at home in the urban than the rural.

This is summed up by Libby Purvis (1996) in her novel about a family torn apart by deviance and sexual infidelity in London being 'healed' by a move to the countryside.[1] The wayward husband reflects on his new rural life following his urban affair:

> London was not after all the whole world. He realised too, that clean winds from the sea and rivers blew over it all, that the air sparkled and the great curve of Norfolk sky gave him new, illimitable thoughts. He could, he thought, at last write his book here. He would not be distracted by the dieselly, soupy, torrid sexiness, the subterranean beat of London. The Norwich girl students, pretty as they were, did not move him particularly and certainly did not disturb him. Such an idea seemed impossible in this clear air. (Purvis, 1996: 83)

As if in response to this dominant portrayal of rural sexuality, recent years have seen the appearance on the market in the UK of 'erotic' calendars of a specifically rural bent. Rural organisations such as the Women's Institute and the S.W. Farmers have produced these 'alternative' calendars providing a *rural* context for the images of scantily clad and naked women and men.[2] Perhaps the clearest message from these calendars is that they do little to contest or destabilise conventional images of rural sexuality. The WI calendar, in particular, produced as 'a bit of fun', simply relies on

---

1 A not uncommon theme in contemporary novels and one that could be illustrated by many other examples.
2 These calendars are now generally available, although we are not able to reproduce them here.

the accepted (and acceptable) images of rural women, coyly or cheekily (semi-)naked. Women are pictured surrounded by their crafts and cakes or in jolly village shows. These women are mainly homely and 'mumsy', they are not aggressively sexual. This is not to dismiss the calendars and their production but simply to suggest that the image we are (supposed to be) most comfortable with in relation to rural sexuality is one of fun and frolics – vaguely innocent and displaying a kind of music-hall lewdness that provides no threat to heterosexuality and does not challenge the place of the family by encouraging promiscuity.

A related but more serious issue here is the association of rape and sexual attack with urban areas. While city streets are perceived as dangerous to women and girls, country lanes are seen as less threatening. Remote countryside may be feared by women due to the lack of surveillance, but the public spaces of the village are believed to be safe. This is partly because of the power of ideas of community, but also due to the fact that the village is not seen as a space for the public display of sexuality.

Another tactic employed to assert the dominance of conventional hegemonic sexualities in rural areas is the comic representation of deviance. Bell and Valentine (1995b) note the treatment of both incest or inbreeding and bestiality in the countryside as a source of humour:

> in English slang, rural dwellers are pejoratively referred to as 'sheep shaggers', while (too) many comedy sketches have traded on the in-bred 'village idiot'. (Bell and Valentine, 1995b: 115)

Homosexual representations

The relative absence of rural homosexual representations is not simply about the assumption of heterosexuality but also concerns the explicit alienation of gays and lesbians from dominant countryside cultures, together with the greater sense of familiarity and belonging experienced by homosexuals living in major metropolitan areas. Binnie (2000) notes the identification of gay men with the city and argues that its visibility within the global city marks queer culture, in particular, as cosmopolitan (see also Oswell, 2000). This cosmopolitanism is based, so Binnie (2000: 166) writes, on 'knowingness and sophistication, a rejection of the provincial and rural'. He suggests that the association of gay men with the city derives in part at least from their mobility (frequently enforced in the face of homophobia). The city is characterised by shifting social relationships in contrast to the stability of the rural community. The representation of homosexual relationships as short term and unstable complies with an image of the city as restless. The city also has the sophistication to embrace and enhance gay relationships in a way that the rural community cannot. Hence it is seen as providing not only a more exciting and fulfilling social and economic existence, but also a sense of anonymity and liberation.

Binnie (2000) also argues that the cosmopolitan representation of gay sexual identity is reinforced by the exclusion of gay men from discourses of nationhood. A major theme of writings on national identity, especially in Britain/England, has been the strong links between ideas of the nation and rurality (see Brace, 1999; Matless, 1998; Mitchell, 2000). Not 'belonging' within the dominant expression of English national identity has served, it has been argued, to marginalise black and ethnic minority groups from the countryside (see Agyeman and Spooner, 1997). According to Binnie, then, alienation from national identity also underpins the exclusion of homosexuals from rurality.

While acknowledging the alienation of homosexual identities from dominant constructions of rurality, Bell (2000a) argues that rurality has been a focus for homosexual imagery. In a paper entitled 'Farm boys and wild men: rurality, masculinity and homosexuality', Bell shifts the focus of debate on the construction of the countryside away from hegemonic representations to consider ways in which gay men have 'sexualized or eroticized' the countryside in their own imaginings, using the countryside, as noted at the start of this section, as a focus or resource for the articulation and performance of homosexual cultures. In so doing he stresses the range of gay imaginings of rurality, showing how each provides a different view of the way cultural representations and lived experiences interweave. It is useful here to look at the different gay rural masculinities identified by Bell (2000a, 2000b; see also Bell and Valentine, 1995b) and to see what particular aspects of rurality they draw on.

One of the most distinct and loaded representations of rural gay masculinity[3] is, according to Bell (2000a), the 'rustic sodomite'. He selects a number of cinematic examples in which the portrayal of rustic sexuality takes the form of homosexual rape. Referring to the cinematic genre 'hillbilly horror', Bell suggests that the connections between masculinity, sexual activity and the rural to be found in films such as 'Deliverance' reflect a 'particularly located cultural construction of US rural folk' (Bell, 2000a, p551). Interestingly, the homosexual masculinity of the hillbillies (matched, it is also argued, in some representations of cowboys and ranch

---

3   Bell himself is not entirely comfortable with this term – see Bell (2000a) for further
    discussion.

hands; see Fellows, 1996) is imbued with a rough, tough rural quality which is pitched against the feminised and effeminate masculinity of gay urban men.

> Men who have faced the 'rigors of nature' . . . [keep] their 'hard riding', 'hard hitting' masculinity intact even when sodomizing each other or any 'sissified' city boys who happen to stumble into their territory. (Bell, 2000a, p552)

Another element of the representation of rural gay masculinity revolves, so Bell (2000a) argues, around the relationship between nature and sexuality. The idea of Arcadia provides, in gay imaginings, a place in which same-sex relationships could be seen as 'natural' (see also Shuttleton, 2000) – a place where gay love and sexual relationships could take place outside the gaze of urban civilisation. Links between rural gay masculinity and nature have been present in a number of social movements through history – the naturist movement in Germany, for example, and the US Fairie movement. Such movements have sought at times to promote a separatist approach to the relationship between gay masculinity and nature, reflecting a 'dissatisfaction with the urban based gay political and social scene'. For Bell this is symptomatic not only of the urban dominance of gay sexual identity but also of the perceived 'take-over' of the countryside by urban gays on a temporary basis for recreational purposes.

As Chapter 3 has illustrated, the association between masculinity, sexuality and nature is not exclusively (or predominantly) homosexual. Bonnett's research on the mythopoetic men's movement in the USA, for example, has shown that nature can also be 'reclaimed' for the purpose of expressing heterosexual masculinity. Nature and wilderness are seen as restorative (again appropriated by *urban* men), allowing men to regain a sense of their 'deep masculinities' which have been eroded by a combination of city life, the feminisation of labour and the power of the women's movement (Bell, 2000a). As with the representation of homosexuality in the 'rural sodomite', the 'back to nature' version of rural gay masculinity is not, it may be argued, particularly powerful in the context of wider representations of rural sexuality. More frequently the perceived deviance of gay and lesbian sexual identities in the countryside elicits the treatment meted out to the women embarking on a lesbian relationship in Joanna Trollop's *A Village Affair* (1989). Here homosexuality emerges to threaten both family and village life and is speedily (albeit painfully) rejected in the interests of stability and the continuation of a rural lifestyle for the children.

This section has considered a number of different representations of rural sexuality and argued that dominant constructions of the countryside, particularly in Britain, serve to reinforce the assumption of heterosexuality and the exclusion of homosexuality. While representations of homosexual relationships do exist, they are generally fetishised, often extreme and almost always deviant. The chapter now moves on from these representations, however, to consider something of the 'lived experiences' of gays and lesbians in the countryside. It examines the ways in which the marginalisation of homosexual identities is reflected in the day-to-day lives of those living in or visiting the rural community.

Homosexual lifestyles in the countryside

Attempting to document the lives of lesbians and gays living in rural communities is problematic due to the related issues of invisibility and marginalisation. There is relatively little academic research on homosexual lifestyles in rural communities, reflecting both the unwillingness of many gays and lesbians in the countryside to 'come out' in relation to their sexuality and the potential dangers faced by homosexuals who may be (directly or indirectly) 'outed' as a result of research. As noted above, the dominance of heterosexuality in both the lived experience and the representation of rural society has fuelled homophobic beliefs that gay and lesbian lifestyles are inappropriate in village communities and encouraged homosexuals from disclosing their sexual identities, on either a temporary or a permanent basis.

The studies that do exist on rural gay and lesbian lives emphasise the isolation felt by homosexual men and women growing up in the countryside (see D'Augelli and Hart, 1987; Fellows, 1996; Kramer, 1995). Jerry Lee Kramer, writing about gay and lesbian experiences in the North Dakota rural town of Minot, notes the problem of seeking out and finding gay and lesbian spaces in rural communities due to the wish for invisibility amongst those homosexuals living in the area. Fear of exposure and of homophobic reaction is, Kramer suggests, heightened in non-metropolitan areas where 'anonymity is a rare commodity' (1995: 209):

> In an urban environment such as Minneapolis St Paul, one has a good chance of encountering homosexuals or homosexuality, even if accidentally. However, in a non-metropolitan environment such as Minot, one doesn't come into contact with openly gay, lesbian or bisexual people

> readily. The gay community is invisible to all but the most diligent searcher.
> (Kramer, 1995: 209)

Similarly D'Augelli and Hart (1987) in a rare study of rural gay and lesbian lives write:

> In most rural areas the gay community is invisible. This invisibility is a result of justifiable fear and discomfort about others' reaction to disclosure of affectional orientation. Many rural gays fear discovery and possible rejection, worrying that any gay behaviour will lead to rejection. Contacts with other gay people may be avoided due to fears of others' reactions.
> (D'Augelli and Hart, 1987: 82, quoted in Bell and Valentine, 1995b)

While one of Jane Cavanagh's respondents in her study of the othering of gay lifestyles in rural communities claimed:

> there are less sorts of facilities to lead a gay lifestyle down here [the rural south-west of England]. So that's another reason why it's not so visible because you can't go into the gay cafes and all of that. You are sort of forced to live a heterosexual life because there is no gay alternative down here. (1999: 97)

Stories of gay and lesbian experiences of rural life identify the problems surrounding access to information, arguing that such problems may be a major factor in persuading rural homosexuals to conceal or deny their sexuality. Indeed Kramer (1995: 208) maintains that the availability and accuracy of local information about homosexuals and homosexuality is possibly 'the greatest difference between metropolitan and non-metropolitan social environments'. He suggests, in relation to his case study of rural North Dakota, that the obvious changes taking place in attitudes towards and experience of the gay and lesbian community of Minot started with the increasing availability of information for homosexuals in public places such as the town library, the university and the bookstore. Other writers (for example Bell and Valentine, 1995b) emphasise the continuing difficulties faced by homosexuals (especially those who have not come out) in acquiring information, particularly that relating to health issues.

Mancoske (1997) identifies the issue of access to HIV/AIDS-related services as a particular problem for rural gays and lesbians. Writing on the incidence of HIV in the United States, Mancoske notes how the early association between cases of HIV/AIDS and urban homosexuality shifted in the early 1990s with the rapid growth of new cases in rural America, particularly the South and Midwest. He goes on to note the problems of caring for people with HIV/AIDS in rural communities due to the poor service provision and lack of funding. In addition, Mancoske cites the 'attitudes and beliefs' of rural residents as a further constraint on the treatment of cases of HIV/AIDS. He writes:

> the delivery of health care social services is influenced by common concerns in rural communities. Various needs and concerns are noted: lack of resources, lack of adequate infrastructure for care delivery, lack of professional providers, geographic barriers to care, a stigmatisation of AIDS, a lack of confidentiality in service provision, lack of psychosocial support, a lack of case management services in rural areas and community attitudes which question the acceptability of receiving services. In the responses to the HIV pandemic in rural America, it is this context which shapes the course of the HIV epidemic. (Mancoske, 1997: 43)

Problems of providing and receiving information about homosexuality clearly come back to the absence of an identifiable gay community in the countryside. Having said this, informal networks that can provide a mechanism for the exchange of information on gay and lesbian issues and services do exist in rural areas – although, by definition, they may be difficult to identify and locate. Additionally, the higher profile afforded to gay and lesbian issues in mainstream journals and newspapers, together with the access to information provided by the Internet and other technologies such as cable television and phone lines, may be helping to address the isolation of homosexuals living in the countryside. As Bell and Valentine (1995b: 117) note, however, such resources may not be available to all and, moreover, suffer from an inability to 'filter tabloid sensationalism and bigotry [putting] even the most basic issues of gay politics and identity' beyond reach.

Poor access to information on homosexuality is exacerbated by the absence of meeting places for gays and lesbians living in rural areas. The clubs, coffee shops, bookshops and other 'safe' spaces for gays to meet in the city rarely exist outside major urban areas and almost never in villages (exceptions are discussed below). Gays and lesbians thus have no obvious place in the rural community to meet potential partners or to gain support and advice from one another. The relatively small number of non-heterosexuals in rural areas,[4] together with the not unrelated wish

---

4   The stated lack of information on gays and lesbians living in rural communities
    means that this is assumed.

to remain anonymous, makes the use of conventional village meeting places (the pub, the village hall, the church hall) impossible and encourages clandestine meetings in places such as cars or public toilets (Kramer, 1995). For those coming to terms with their gay identity the lack of a definable homosexual space in rural areas may be particularly inhibiting and may increase the likelihood of gays and lesbians continuing to conceal their sexual identities.

For many rural gays the obvious response to the isolation and ignorance of countryside living is to abandon it, either temporarily or permanently, for the city. There is, as writers such as Binnie (2000), Oswell (2000) and Weston (1995) argue, a strong relationship between homosexuality and the urban, not only in terms of representations of gay and lesbian identities, but also in the practicalities of daily life. As Binnie (2000: 172) states, 'coming out and developing a gay identity has commonly gone hand in hand with becoming a sophisticated urban dweller at ease with urban life'. Weston (1995, quoted in Bell, 2000a) writes, similarly, of the city as a 'beacon of tolerance' for gays while the country is a 'locus of persecution and gay absence'. Bell and Valentine conclude from a review of studies of rural gay and lesbian lives that the

> isolation, unsupportive social environments and chronic lack of structural services and facilities lead to eventual or projected emigration to larger (urban) settlements which offer better opportunities for living out the 'gay life'. (1995b: 116)

For those who decide that permanent residence in the city is the only or preferred way to negotiate their sexuality, there may be a painful break with family and friends. Others may not have the desire or option to move on a permanent basis and may as a result be restricted to periodic visits to the city to pursue sexual relationships and other friendships. Such visits are dependent on access to transport (often problematic for the young), time and financial resources while, as Kramer (1995) points out, they can in some areas be subject to weather conditions.

Throughout this chapter reference has been made to the homophobia experienced by gays and lesbians living in rural areas. As well as fundamentally shaping the representation of homosexuality and gay relationships in the countryside, such homophobia also has very practical repercussions for the everyday lives of gays and lesbians in rural communities. Some studies have described the hostility faced by rural homosexuals on a daily basis (see Cavanagh, 1999) which, at its extreme, may deny them access to spaces within the village – the most obvious being the pub (see Watkins,

1998) – or to village events. Fear of abuse or of simply not being accepted within the rural community may lead many gays and lesbians to conceal their homosexuality – with significant effects on their social life. The dangers of being 'outed' are seen as potentially more far reaching in a tight-knit rural community where difference of any sort rarely goes unnoticed. Bell and Valentine (1995b: 116) suggest that the 'intense heteronormative pressures of rural life' mean that many rural gays and lesbians are married or have long-term relationships with people of the opposite sex. While precise figures are clearly impossible to obtain, studies of the lives of rural gays and lesbians claim that the chances of remaining in an unsatisfactory relationship are greater for those living in the countryside due to the heightened fears of discovery and homophobic reactions on the part of other villagers.

> Anti-gay violence remains a dominant fear for rural lesbians and gay men, . . . further restricting the opportunities for social, political and sexual liaisons. (Bell and Valentine, 1995b: 117)

Gay visitors to the countryside, it is reported (see Cavanagh, 1999), may also experience hostile and homophobic attitudes from 'host' communities. While levels of information and support available to gay and lesbian visitors to the countryside have grown over recent years, homosexuals may still find themselves singled out by rural residents unused to encountering anyone not of heterosexual orientation. As temporary visitors, however, the alienation from rural society may be less significant to broader attempts to make sense of the countryside and to their own lives within it. As noted above, homosexuality is frequently associated with the urban and as such has no place in the countryside. Gay male visitors, for example, may be passed off as effeminate, sissy urbanites who do not share the 'real masculinity' of (heterosexual) country men. The unnaturalness of gays and lesbians alike signifies that they are city folk and are distanced from the rural environment. Occasionally gay and lesbian visitors to the countryside are criticised by homosexuals living in rural areas in the way that they respond to anti-gay attitudes. The desire to 'fit in' often makes those gays and lesbians living in the countryside more conscious of the need for caution in responding to hostility and homophobic behaviour.

Thus far this review of work on the lives of gay men and lesbians living in (and visiting) the countryside has painted a somewhat disturbing picture of exclusion, isolation and violence. Rural communities, it has been argued, can provide very negative experiences of being gay and while homophobia is an issue for rural *and* urban environments, the particular characteristics of rural society – the dominance of the family and the size and tight-knit nature of the community – heighten the existence of anti-gay attitudes and create a greater sense of exclusion. It must be recognised, however, that some more positive experiences of rural gay lives can be found (see Bell, 2000a; Fellows, 1996) and it would be wrong to move on without acknowledging the nature and basis of these experiences.

A number of accounts of positive gay and lesbian experiences of rural living draw on ideas of nature and on the relationship between natural-ness in various forms and homosexuality (see Bell, 2000a; Fellows, 1996; Valentine, 1997). The countryside can be seen, for example, to offer free-dom from the 'controls' of urban life and from society's homophobic tendencies. There is space in the countryside, so it is argued, to choose the way you want to live with less interference from the 'unnatural' environment of the city. Bell (2000a), reviewing the study of gay farm boys by Will Fellows (1996), notes the sense in which the countryside can provide an environment in which gay men and boys can more success-fully 'avoid peer pressure and invent themselves according to their own inclinations and standards' (Fellows, 1996: 16, quoted in Bell, 2000a).

Fellows (1996) recounts the stories of rural gay men who grew up in the Midwest of America.[5] Amongst these stories are a number of positive experiences of rural living, with respondents talking about being in touch with nature, having space to develop their own identity and realising the value of hard manual work.

One of the respondents in Fellows' research describes his own situation growing up on a farm in Iowa in the 1940s as follows:

> Growing up close to nature, close to all those things that you see come to life, gives you a completely different perspective on how you deal with other people as well as yourself. Seeing life become life, respecting that, watching that happen, watching things grow – I kind of did the same thing with myself. It was totally uncluttered. I didn't have to deal with lots of people. (Fellows, 1996: 82)

As Fellows (1996) himself warns, however, while recognising some of the positive experiences of rural gays, it is important that we do not allow the idealised imagery of the gay urban dweller to cloud the reality of what it is like to be a gay in the countryside. He quotes from Silverstein's work on gay couples in America:

---

5 Will Fellows conducted interviews with 75 men from rural farming backgrounds in the USA during 1992/3. Their stories form the basis of Fellows' book, *Farm Boys: lives of gay men from the rural Midwest* (1996, University of Wisconsin Press).

> City gays imagine the boys on the farm as somehow more wholesome than themselves. Soaking up the sun while pitching a bale of hay, their bodies taking on a bronze glow, these promising young men develop tight muscles from manual labour and hardiness; the lines on their faces and the callouses on their hands are the results of wind, rain and the warming sun. In short, they are pictured as country bumpkins with rosy cheek, ready to be plucked if they venture into the big city. (Silverstein, 1981: 241, quoted in Fellows, 1996)

While one of Fellows' own respondents suggested that:

> 'A lot of men idealize the naïve, good-looking tanned farm boy.' (Fellows, 1996: xi)

Gill Valentine (1997) identifies a sense of space and freedom in the positive attitudes of rural lesbians in the United States. The creation of separatist communities in the countryside (as mentioned in Chapter 4 in relation to alternative discourses of community) was a feature of the lesbian and gay movements in 1970s America. Living in rural areas allowed lesbians[6] to distance themselves from mainstream society and establish 'ideal' communities away from the influence of men and from patriarchal social relations. The space of rural areas afforded women the freedom and isolation to remain 'purer in their practices' (Valentine, 1997: 111), to grow their own food and set their own rules. The view was that nature 'protected' them and that land ownership allowed them control over their own space. This was essential, as Valentine notes, because it:

> [gave] women the freedom to articulate a lesbian feminist identity, to create new ways of living and to work out new ways of relating to the environment. (1997: 111)

These positive reflections on the countryside by gays and lesbians are interesting. They do, however, relate primarily to experiences in *remote* countryside. The separatism of the Lesbian Lands described by Valentine is significant and suggests, as in the case of the cowboys on the range, that the ability to get away not only from urban society but from others living in the countryside is critical. The freedom experienced by the subjects of these examples results from their ability to avoid contact with anti-gay attitudes. Examples of gays and lesbians living happily in the heart

---

6   Separatist rural communities were created by both lesbians and gay men but were mainly a radical lesbian strategy for escaping patriarchal structures and relations within society.

of the traditional village, expressing their sexual identity freely within the rural community, are less easy to find.

The idea of the natural environment is central in both Fellows' and Valentine's stories of gay men and lesbians in the countryside. Both, however, provide a different interpretation of the relationship between sexuality and nature. In Fellows' examples, the gay masculinity of the farm boys is described as tough and 'butch'. These men see themselves as 'hard' men and one of their main concerns about their sexuality is that they should not be labelled as 'sissies' or 'fairies'. Again, as Bell (2000a) notes, the 'farm boys' often subscribed to a view that homosexuality was urban – 'an unnatural phenomenon of the city that had no relevance to rural life' (Fellows, 1996: 18, quoted in Bell). Their exaggerated butchness was designed to demonstrate their rural authenticity as well as conceal their sexual identities.

> In their performance of gendered and sexualised self-identities, then, men tended towards 'butchness', in part as a way of covering over hints of homosexuality, and often in part as a reflection of their subscription to negative views of urban gay effeminacy. (Bell, 2000a, p556)

This view of conquering or dominating nature as indicative of hetero-sexual or 'real' masculinity is also noted by Woodward (1998) in her work on military training (as discussed in Chapter 3). She makes the point that soldiers who 'fail' fitness tests in the countryside are seen as 'less of a man' by other soldiers. She writes:

> Again and again in his account of the selection procedure, Ballinger [a former SAS soldier] records how a failure to achieve dominance over the landscape by finishing training exercises on the hills is equated with effem-inacy; failures are called Girl Guides. The aggressive heterosexuality and homophobia often identified with this construction of hegemonic mas-culinity also comes through; failures are labelled 'queers' and 'fairies'. (Woodward, 1998: 288)

The interpretation of nature (especially wilderness) as a male space in which 'real' masculinity can be asserted contrasts strongly with the lesbian experiences of nature discussed by Valentine (1997). Closeness to nature was seen as an important principle of the Lesbian Lands and one that drew on essentialist notions about women's relationship with Mother Earth. As discussed in Chapter 3, men were portrayed as the destroyers of nature, women as the carers and nurturers who could protect (and be protected by) the natural environment. The lesbian communities aimed at being self-sufficient and developing skills of food-growing and other crafts, again emphasising the links between sexual identity, freedom and rurality. Many of the Lesbian Lands also included a spiritual dimension, celebrating events and times associated with nature and the earth – the full moon, equinox and solstice. Valentine (1997) quotes Cheney, who documented the Lesbian Lands, drawing extensively on the voices of women living in (or who had lived in) the lesbian rural communities:

> 'We view our maintaining lesbian space and protecting these acres from the rape of man and his chemicals as a political act of active resistance. Struggling with each other to work through our patriarchal conditioning, and attempting to work and live together in harmony with each other and nature.' (resident of Wisconsin Womyn's Land Cooperative, in Cheney, 1985: 132)

What is interesting about the lived experiences of women within these lesbian separatist communities is their appropriation of (a version of) the rural idyll for their own particular purposes. The communities were formed with a vision of a rural Arcadia, to provide a space for women (lesbian and straight) to express their identities freely in an idealised natural and social environment, distanced from male power and control. Ultimately, however, as Valentine (1997) reports, the communities fragmented and divided over issues such as land ownership, class and separatism.

> Thus, lesbian separatist attempts to establish 'idyllic' ways of living in the countryside appear to have unravelled because, in common with traditional, white middle class versions of 'rural community', attempts to create unity and common ways of living also produced boundaries and exclusions. (Valentine, 1997: 118–19)

Conclusion

This chapter has attempted to draw together a number of ideas raised in previous chapters on a particular element of gender identity, sexuality. As Chapter 2 has discussed, geographers have looked increasingly to debates on the construction, reproduction and performance of sexuality in attempting to understand social, cultural, political and economic transformations (Binnie and Valentine, 1999). Moreover, those from outside geography have started to foreground the study of space and place in their examinations of sexuality. Despite this growing interest in the relationship between sexuality and spatial patterns and processes, a direct focus on sexuality has been relatively slow to influence rural studies. This is not really surprising given the more traditional history of rural geography, identified in Chapter 2; it does, however, make any attempt to summarise past research on the topic somewhat limited.

The relative scarcity of past work should not, however, be allowed to obscure the potential richness of the rural as a site for the discussion of sexuality. As I have tried to show in this chapter, a focus on sexuality can contribute greatly not only to the study of gender relations and marginality within rural society, but also to attempts to understand more about the cultural construction of rurality and its relevance to the lives of those living in the countryside. To some extent there is a need for rural geography to 'catch up' with the rest of human geography in mapping the relationship between sexuality and space in the countryside and examining the material and the everyday – how the experiences of gays and lesbians are lived out in rural communities. As this chapter has pointed out, the village is not often portrayed as a sexed space in the way that parts of the city have been. The village is a space for families and for contained and predictable sexual relations; something that is highly relevant to the construction and reproduction of social and gender relations more broadly with rural society. The focus on sexuality here has highlighted the current interest amongst geographers in the body. Again, the rural provides a rich source for the discussion of the relationship between masculinity, femininity, embodiment and the environment. The chapter has shown how the rural is eroticised through the positioning of the gay male body in the rural landscape, as closer to nature. Similarly, the lesbian body is also portrayed as 'purer' through its rejection of the power of patriarchal masculinity and, in particular, the sexed male body as 'apart from' nature. It is the view of the gay and lesbian body as threatening and predatory and the heterosexual body as safe that constitutes, perhaps, the most common assumption of rural embodiment. Despite being supposedly 'closer to nature', rural areas are not associated with highly explicit celebrations of the body, particularly celebrations that go beyond what might be considered by many to be 'normal' expressions of heterosexuality. Bell and Valentine (1995b) cite, for example, the heavy-handedness of police action in breaking up a 'leather party' in the Yorkshire village of Hoylandswaine in England. As noted in the chapter, the sexed rural body is either a figure of fun, represented as rather coy or cheeky, or it is disturbing and 'unnatural', to be hidden or ignored.

One of the clearest messages of this chapter is the continuing need for research on the lesbian and gay geographies of rural communities and spaces. As Binnie and Valentine (1999) point out, some progress has been made in the adoption of a specifically rural focus to debates on the experience of homosexuality; rural gays and lesbians being one of the groups of 'neglected others' identified by Chris Philo (1992) as not only potentially living a very different rural existence to the habitually studied, Mr Average, but as such, deserving of attention by rural researchers. Yet there remain many aspects of rural sexuality that have received scant attention. In addressing this continuing neglect it is important, however, that we also go beyond the simple identification of sexuality in rural places and look at the relationship between sexuality and processes of social change in the countryside. We need to understand the rural as a predominantly straight space and to think about this not only in the way that it affects the lives of gays and lesbians but in its broader implications for the restructuring of rural society.

8 | Conclusion

The continuing relevance of gender in researching the rural landscape and society?

It may seem strange, having reached the final chapter of a book on gender and rural geography, to start to question the legitimacy of gender as a defining category for the interrogation of rural social relations. So far I have examined in the various chapters the different perspectives adopted by geographers in the study of gender difference, showing how these contribute to the understanding of rural society and community, but have not disputed (or defended) the relevance of a gender perspective more generally. In this final short chapter, however, I want to step back from the detail of previous chapters to consider the continuing value of debates on rural gender divisions. I do this essentially with the intention not of discarding the previous work but rather of reasserting the importance of a focus on gender in rural geography. Hopefully this discussion will not only situate rural gender studies within very contemporary debates in feminist geography but will help to identify possible directions for future research.

The work on gender and difference discussed in Chapter 2 has helped to frame recent debate in feminist geography and has provided inspiration for new research on aspects of masculinity and femininity and differences in the articulation and performance of gender. There was a time when the recognition of hybridity threatened to undermine the value of the very categories 'male' and 'female' for the purposes of academic enquiry (Pratt, 1999), but as theoretical debate shifted to the examination of constructions of masculinity and femininity and to the exploration of identity, the legitimacy of a focus on gender was sustained. In rural geography, as this book has attempted to show, work has tended to cling on to conventional gender categories and been less certain about embracing the notion of gender difference, either theoretically or empirically. It is perhaps this lack of work on the differences between rural women that has impeded a broader focus on rural gender identity.

The rather slow progress of rural gender studies is consistent with the more general 'lag effect' that has surrounded the take-up of theoretical developments in geography by those engaged in rural studies. However, I suggest it is not entirely about a reluctance to confront new perspectives but also about the particular characteristics of rural gender identities. What has become apparent through the different chapters of this book is, I hope, the widespread existence of a dominant cultural construction of rural womanhood in developed capitalist economies which both shapes and is shaped by the endurance of highly traditional rural gender relations. This is not to say, of course, that difference between rural women does not exist. It is, however, to recognise the power of a particular version of femininity within rural society and to appreciate the exclusivity of that version within contemporary rural gender relations. What is so strong in the rural context is the relationship between gender identity and place, in particular the belief that a traditional construction of womanhood is more appropriate to rural society. The link between patterns of social relations and community interaction in rural areas is seen to depend on and reinforce the conventional identities of women as wives and mothers. Thus in some senses the articulation of difference amongst rural women is somewhat alien since women who are 'different' are not truly 'rural'.

This dominance of a particular form of gender identity for rural women has been evident in recent work on rural marginalisation. The focus on rural women as one of the so-called 'neglected rural others' has demonstrated the lack of tolerance surrounding the construction of rural femininity and the persistence of traditional assumptions about the gender identity of rural women. Work on sexuality and ethnicity has highlighted the problems faced by lesbian women and gay men and by black and ethnic minority people. It has also demonstrated the power of constructions of rurality and of the dominance of 'the Same' – those who conform to accepted rural identities. Women who comply with these accepted identities are not marginalised from rural life or from the rural community; they may, however, be 'othered' in relation to different versions of femininity.

The continued importance of a specific and highly traditional version of rural femininity is nowhere as obvious as in relation to childcare. Recent research by Halliday (see Halliday, 1997; Halliday and Little, in press) has shown how the mothering role of women remains dominant (and often exclusive) both practically and ideologically for rural women. In this context there is a very real need for continued emphasis on gender in academic research, as a defining feature of rural social relations. How different women manage their lives and juggle the demands of domestic and paid work is clearly of considerable relevance to their day-to-day experiences, constraints and opportunities, but, for the majority, how they make sense of their rural lifestyles and, in particular, their place in the rural community, is in relation to their roles as mothers. This also applies to women who are not mothers (and grandmothers), whose presence in the family space of the rural community may be questioned (see Little and Austin, 1996).

While rural women's identities as wives and mothers have been the subject of some (albeit rather limited) research, very little attention has been given to men's rural gender identities, except in terms of the relationship between masculinity and nature. We know very little about men's identities within the rural community and how these are worked out in their daily lives. The highly conventional nature of women's domestic roles, however, presupposes equally traditional masculine gender identities within the rural family. Indeed, Halliday's research on rural childcare, cited earlier, notes the long hours spent working (and travelling to work) by rural fathers and their consequent absence from the home at key times in their child's day (such as mealtimes and bedtime). Other research has noted that men's involvement in the rural community tends to be in relation to sporting events rather than in the organisation of children's activities (Jones, 2000; Little, 1997b). Such research is not particularly detailed, however, and the examination of masculine gender identities has not generally been the main focus. There is a clear need for research to identify the nature of men's roles within the rural household and community and to place these within an examination of rural gender identities. Changes in the patterns of employment, for example, with more people working from home, may be having an effect on the gender division of labour within the rural household.

Having made a case for the continued relevance of gender as a defining social category, it is important to recognise that the resilience of stereotypical gender identities amongst rural women and men does not negate a consideration of gender difference. As noted, certain gender identities may be 'othered' by dominant constructions of rural womanhood, yet these constructions themselves also incorporate variety. In other words, women of various ages, classes and backgrounds can 'buy into' the overarching identity of rural women as mothers and wives. In so doing, however, their experiences as rural mothers and wives may vary considerably. In the chapter on employment, for example, I argue that women on low incomes or caught in the 'benefits trap' will have very different opportunities and lifestyles from those middle-class women with professional employment and their own car. In the same way Halliday's (1997) research has shown that the childcare options available to rural women with extended local family networks may transform their ability to access the labour market.

Research on the geographies of rural young people (see Jones, 1999; Leyshon, 2000; Little and Leyshon, 1998) has begun to show how the rural lifestyles and attitudes to the rural community vary between generations. The work does, however, suggest that while young people may resist the expectations placed on them by adults in the countryside and resent the control over their use of rural space, they also buy into many of the dominant ideas surrounding gender identities in the countryside. There is evidence to suggest, for example, that rural youth envisage leaving the countryside to pursue careers and gain new experiences, but that they also see themselves returning to rural communities to 'settle down' and have families. Despite often feeling bored and restricted in their home villages, many still believe that the countryside is a 'better' place to bring up children and to get the most out of family life. These sentiments incorporate firm ideas about the roles of men and women in the countryside and point to very conventional ideas about rural gender identities. These contrast with gender identities believed to exist amongst 'townies'. Leyshon's research, for example, has revealed constructions of urban femininity amongst rural young people that are very different from those of rural femininity; his work showing how young men and women living in rural communities referred to sexually active, young, urban women as 'slags' and 'sluts'.

In documenting the lives of rural people and investigating the different ways in which gender identities are constructed, negotiated and performed, we need to think about how research is undertaken and data collected. Issues surrounding personal lifestyles and beliefs are clearly sensitive by their nature and require the adoption of appropriate research methods which will not only allow the collection of accurate information but will do so in a way that will not compromise the position of the subject or the relationship between the researcher and the individual/ community being investigated. This next section will consider the ways in which research on rural lifestyles and gender identities may be carried out and, in particular, the construction and adoption of feminist methodologies for researching rural gender issues.

Gender identity and research methodology

Geography's much-discussed 'cultural turn' has, amongst other things, encouraged the adoption of qualitative research methods and, in particular, the use of ethnographic techniques of data collection (see Anderson and Gale, 1992; Cook and Crang, 1995; Cook *et al.*, 2000). Clearly, ethnographic methods have a long history of use amongst anthropologists and sociologists, and even within geography their application in a variety of fields predated the rise of the cultural perspective (see Jackson, 1983). However, the attention to difference, to lifestyle and to the individual which is central to research in cultural geography, demanded that such methods be more widely applied and developed according to the particular direction of geographical enquiry. Thus one of the results of the penetration of the 'cultural turn' into rural geographical research has been a greater engagement with ethnographic research methods including participant observation, focus groups and the use of textual, photographic and filmic data sources (see Bennett, 2000; Goss, 1996; Revill and Seymour, 2000).

Rural social scientists are not, however, new to the application of ethnographic research methods. The era of the rural 'community studies' (discussed in Chapter 4) saw the use of, in particular, participant observation in the collection of research data – indeed Phillips (2000: 28) argues that the use of ethnographic techniques in more contemporary rural geography 'could even be seen to be a return to the techniques used by earlier generations of rural researchers'. He suggests, moreover, that we have much to learn from the discussion of 'community studies as method' (see Bell and Newby, 1971). Some of the description from the rural community studies of the 1960s and 1970s reads rather like a report from the far-flung parts of other continents as researchers immersed themselves in the ethnographic detail of rural societies. Rarely do such studies articulate the detail of the methodologies adopted, however, and while researchers are clearly living within their chosen communities they also appear detached from them. As community studies went out of fashion so there was a decline in the use of ethnographic methods in rural studies while in rural geography, in particular, the preference for descriptive, land use-based research encouraged the adoption of large-scale surveys and quantitative methodologies.

While the 1970s and 1980s saw rural geographers involved in some detailed studies of rural people and lifestyles in, for example, research on farm families or on transport and housing issues, much work at the time focused on structural processes, patterns of economic restructuring and political power. Such research often incorporated interviews with actors in key positions of authority – landowners, planners and other political and economic elites – and even where research included the examination of the effects of political economic processes of transformation on the lives of rural people, it rarely employed more ethnographically based methodologies in so doing. Even research on social and welfare-based issues such as rural deprivation was based on an (often quantitative) examination of the material conditions of people's lives – their income levels, labour market situation, housing provision and access to a car – rather than on more experiential aspects of their rural lifestyles (see Cloke *et al.*, 1994; McLaughlin, 1989).

It was not, then, until recent interest in the differing ways in which people experienced the conditions of rural life, especially those felt to be marginalised by mainstream expectations of rural society, that rural researchers turned once more to ethnographic methods in any conscious or significant way. As Hughes *et al.* (2000: 5) suggest, it is this permeation into rural geography of the 'epistemological reflection engendered by the engagement with contemporary critical thinking' that has encouraged the use of a range of more qualitative research methods. In examining the lives of different rural residents, researchers have needed to be sensitive not only to the particular identities of those being researched (especially where those identities may attract hostility in society generally – for example gays and lesbians, travellers, black and ethnic minorities), but also to the ways in which those identities interact with rurality – in other words, to the ways in which rurality imposes a particular set of constraints and expectations on the performance of identity. The methods through which the lives of rural people are studied must be alive to and able to accommodate this sensitivity (see Cloke and Little, 1997).

What is particularly interesting for the purposes of this chapter is the link between ethnographic and feminist research methods. Ethnography as a method of data collection represents, in its sensitivity to difference, culture and to individual experience, as noted above, a valuable tool for researching the lives of rural women and gender identities. This sensitivity in the process of data collection is one of the qualities called for by feminist geographers in their critique of research methods in the 1980s. Monk and Hanson (1982), for example, criticise much research in human geography for failing to identify appropriate questions for establishing women's needs and for gender-blindness in the selection of both variables and respondents (see also Roberts, 1981). Their call for research to adopt non-sexist methods, sensitive to gender issues, has been echoed and built on by feminist researchers in geography ever since the early 1980s.

Methods such as ethnography and open-ended interviewing have been identified as offering an alternative to the positivist methods which form, so feminists argued, part of the patriarchal production of knowledge (Dowler, 1999).[1]

Whether or not it is possible to identify a specifically feminist research method or set of methods has been the focus of contentious debate amongst social scientists (see Dowler, 1999; McDowell, 1992). This debate, while perhaps not resolvable in itself, has highlighted the distinction between methods as *tools* for research and methodology as *theory* about the research process. Feminists thus may contest the existence of a feminist research method but are agreed that what are needed are techniques that 'are consistent with their feminist convictions' (Dowler, 1999: 93). As McDowell argues:

> there is a broad agreement that feminists, within and outside their own discipline, are searching for methods that are consonant with their values and aims as feminists, and appropriate to feminist topics. (1992: 405)

She notes the difficulties involved in deciding whether particular methods are any more appropriate than others for conducting feminist research but suggests that feminist researchers insist in general on more collaborative and non-exploitative methods than are used in 'conventional' research.

McDowell (1992) talks in particular about the power relations involved in the research process and the importance attached by feminist researchers to the need to break down 'the typically unequal power relations between a researcher and her informants' (McDowell, 1992: 405). This, she suggests, can best be achieved through the adoption of qualitative, detailed case-study work which allows interviewees to have some ownership of the research process. In this way the researcher is not seen as a distant, disembodied and rational figure but as a sympathetic equal. Interviews thus become more of an interactive process in which there is a dialogue between the researcher and the subject (see also Bennett, 2000). Feminist geographers have written about the strategies that can be adopted to ensure that the methods they use promote collaboration – thinking about the ways respondents are selected and approached, the environment in which the research takes place, the language used etc. – but the point is

---

1   Some feminist social scientists have pointed out that quantitative methods *per se* should not be dismissed and that they can perform an important role in feminist geographic research (see McLafferty, 1995). As Dowler (1999) notes, many feminist geographers are now combining the use of qualitative and quantitative methods in research.

made that it is only through commitment within the research process as a whole that such strategies can achieve a more collaborative outcome. Skeggs (1994) is more sceptical about the ability of any research, feminist or otherwise, to break down hierarchical power relations in the research process. She suggests, however, that through the inclusion of reciprocity in the construction and implementation of research methods, the power of the researcher can be used to empower the researchee. Skeggs cites Oakley (1981) in arguing that 'researchers should productively use their power by giving any information and knowledge that they have that may be useful to the researched' (1994: 82).

As well as recognising the power relations involved in the research process, feminist epistemology stresses, as the Women and Geography Study Group (1997: 87) points out, the non-neutrality of the researcher. The Group contests the boundaries between 'fieldwork' and everyday life, echoing Katz's belief that 'we are always in the field'. Feminist geographers have argued that the 'positionality' of the researcher is relevant not only in terms of power relations but more broadly in the approach to the subject and in the interpretation of data. As the WGSG writes:

> the personal affects the way in which we do research: it influences the questions we ask, the ways in which we interpret answers to those questions and what we do with our research results. (1997: 88)

Rural geographers have also begun to recognise the subject position of the researcher as relevant to the research direction and outcome (see Cloke, 1994; Fielding, 2000). Martin Phillips (2000) shows how rural researchers have explored, and responded to, two specific issues highlighted by the notion of positionality. First is the acknowledgement, as noted by feminist researchers above, that research involves power and that, drawing on the ideas of Foucault, knowledge can be seen as a source of power. Thus rural researchers have had to question not only how but also why they are undertaking research and to step back from a desire to influence the world and think about how their research is 'connected into relations with others' (Phillips, 2000: 31). Secondly, rural geographers have had to recognise that their own 'position' cannot be separated from the research process. Phillips draws on standpoint theory to highlight the importance of positionality but then goes on to question whether we can ever fully address the issues and problems that it raises. It is not enough, as he points out, to simply declare our 'position' in respect to, for example, gender, ethnicity, class – to see, in effect, positionality as merely the statement of personal biography – but we must seek to understand how the intersubjectivity of research affects its very meanings, purpose and boundaries.

To apply these ideas and worries to rural gender research I (re)turn, very briefly, to my own experiences of researching rural women. Most of this research has been based on either questionnaire surveys or interviews with rural women in the UK about their employment experiences and their roles in the rural community. In many ways my 'positionality' within the research process should present me with relatively few problems in this particular research context – as a middle-class white woman interviewing other mainly middle-class white women some of the most obvious pitfalls are avoided. A past difficulty of seeming much younger than many of those I was interviewing seems, strangely, to be receding! I have also lived in a rural area and have always had a strong interest in the countryside – I enjoy walking and other 'rural' recreational pursuits, for example. I think I also start from an advantage shared by many fellow 'rural' researchers, namely that rural residents often seem to have a keen interest in talking about 'their village' and show a concern with and about any study of local people and places. The smallness of rural communities and the efficiency of local networks of information and gossip can make 'finding out' a relatively painless business.

However, I have also been struck during the course of my various pieces of rural research with regard to how 'other' I am to the women I am studying. While generally happy to take part in research and, in many cases, seemingly genuinely interested in the 'purpose' of the research, the women are frequently very distanced from its underlying direction and motivation. This comes back to the strength of the domestic ideology and the belief, amongst many of the women I talk to, of the essential 'rightness' of their gender roles. Any suggestion to the women that they may be 'marginalised' or disadvantaged by the constraints of their domestic role would, to many, be offensive or simply inaccurate (a finding echoed by Cavanagh (1999) in her research). Responses of this kind have been 'managed' (by both the respondents and myself) as a function of my lack of understanding of rural life. The gulf between myself and the subjects of my research has seemed, at times, immense – interestingly, a gulf that the rural women were only too happy to imagine as reconfirming a view of the separatedness of country and city and the association of academic work with the world of urban progress and consequently alien to the concerns of the countryside.

This gulf has at times fundamentally shaken the belief I have in my research and in how I am conducting it. In particular it has compromised plans I have had for sharing research findings and constructing a 'research alliance' between myself and my research subjects. I have questioned, at times, whether I am guilty of thinking I 'know best' for the rural women I am studying in relation to both the day-to-day operation of their own lives and the wider operation of gender inequality. The research, for example, on women's employment has unavoidably (given my beliefs) taken the standpoint that women should have certain choices in relation to paid work – particularly whether they do it and what they do – and that those who do not have this choice are, to some extent, disadvantaged. I have not denied that this disadvantage varies between women and over time and that it may be a 'price' that some are prepared to pay. I have, however, been accused of undermining the role of women's domestic work and in assuming employment to be somehow superior to staying at home. I hope this is not the case but I can see how it may be a natural assumption from the research I do and the questions I ask. I also hope that I have tried to take these issues on board in designing subsequent research, although recognising that I cannot completely revise my basic standpoint.

The big question that remains is whether we can or should attempt to research the lives of those who are very distant from us. There is no doubt that in certain situations a sensitivity to the research topic and/or the respondent may dictate aspects of the researcher's 'positionality' – the obvious example being that the interviewing of women who have experienced sexual harassment should be done by women. But there are many situations that are not as clear-cut or apparently as sensitive; should, as has been suggested to me, only women with children do research on the choices and problems around childcare? Clearly at one level such attention to common interests and positions is unsustainable and would, if pushed to its obvious conclusion, render all research inappropriate. However, thinking about the issues raised by questions of positionality can help to increase sensitivity within the research process and minimise the inequality of power relations. What is important is perhaps not that the subject positions of researcher and respondent are identical but that strategies for data collection and for dissemination of research findings are sympathetic to the particular characteristics, experiences and needs of the subject.

Further research

Many of the chapters of this book have concluded with an acknowledgement of the gaps in our understanding and ideas for future research. Here, by way of a final concluding section, I want to draw these research threads together and highlight what I see as key topics and directions for future studies in gender and rural geography. It is, I believe, an exciting time for work of this nature. Gender is now an accepted and legitimate research interest within rural geography. It remains, however, relatively under-developed, in terms of both actual data gathering and original research findings and the generation and discussion of more theoretical and conceptual ideas. At the same time there has been some fascinating work on gender within the broader disciplinary context and this has opened up new questions and possibilities for a rural focus. It is thus an excellent time to take stock of some of the important concerns within past and contemporary rural gender studies and to develop this work in the context of more recent debates on gender identities, the body and sexuality and space.

While this book has taken 'gender' as its focus, relating geographical work on gender role, gender relations and gender identities to the rural, it has done so largely through the lens of women and femininity. This, in the present context, is defensible but rural research specifically on men's lives and the changing relationship between men and women's ruralities is, as noted above, urgently required. We need a much more detailed understanding of the negotiation and management of gender relations and of the intersection between men and women's experiences of the rural. A limited amount of work on masculinity has raised interesting ideas about the expectations and performance of male gender identities in rural areas – again, this work needs to be developed and extended in examining not only the marginalisation of, for example, gay masculinities, but also more central and mainstream expectations of rural masculinity (see Evans, 1999). Thus while existing work has explored cultures of rural womanhood and femininity in the context of the rural family and community, there has been almost no work on the relationship between constructions of masculinity and rural society.

Work on sexuality and on youth has started to draw attention to differences in the experience and performance of gender identity amongst women and men as well as simply between them; this work, however, has not gone beyond the articulation of more major divisions. This is partly because of the adherence, as noted above, to certain key characteristics of rural gender identity and the apparent consensus surrounding notions of rural femininity and masculinity. The perceived dominance of these mainstream constructions should not, however, obscure their contestation since they are important not only in themselves and their alternative 'take' on gender and rurality, but also in their power to reinforce accepted versions of gender identity. Future research needs to move away from the idea of gender as fixed and stable and to start to recognise the fluidity and uncertainty in the articulation of gendered identities in a range of different situations. Such research would allow a greater appreciation of the fragmentary nature of experience within the rural community and start to explore the possibilities of shifts even in the more traditional elements of rural gender identity.

Beyond the more general study of gender identity, one of the most exciting opportunities for the development of rural gender studies concerns the geography of embodiment. Again this is an area which has received little research attention and yet its relevance to many of the interests and concerns of rural geographers cannot be overstated. Clearly the body, how it is constructed and imagined in a rural context, is critical to the wider theoretical debates surrounding gender identity, femininity and masculinity, as we have started to see in work on nature, landscape and the rural environment. Within these debates, however, the centrality of the body to social relations within the village, to the expectations and assumptions surrounding gender relations, has gone largely unrecognised. There is a need for research on the embodied nature of social change within the rural community and for a consideration of the importance of the body in debates on issues such as rural governance, the rural labour market and tourism and recreation. While the spatialities of the rural community and landscape have been examined in relation to, for example, children's geographies (Jones, 1997, 1999) little consideration has, as yet, been given to the relationship between the body and the use of space in the countryside.

More specific and established topics within rural geographical research also open up the possibilities for a gender perspective. Recent work on technology and its role in the changing employment patterns and lifestyles of rural residents, for example, has clear relevance to debates surrounding the gender division of labour within the home and the community (Wheelock et al., 1999), and indeed for the wider relationship between gender and rural society. The continuing restructuring of agriculture and the problems faced by farm businesses in many capitalist economies raise a new set of questions concerning the roles of different members of the farm household and the cohesion and survival of the farm family.

The scope for continuing research on gender and rural geography is considerable and this short discussion has mentioned just a few possible research avenues for the future. It is not the purpose here, however, to construct a detailed list of potential topics but rather to emphasise the continuing and expanding opportunities for research that develops the theoretical and conceptual underpinning of feminist rural geography and that provides detail on the gendered nature of patterns of social and economic change in rural communities. While singling out here some ideas for confronting gender issues in rural research, it is also important that we do not see gender as occupying a separate research agenda. It is essential that we continue to heed the well-worn concerns that have long occupied feminist geographers – that as well as giving space to specific and highly focused research on gender relations, gender identities and the inequalities existing in men and women's lives, we ensure that the consideration and awareness of the importance of gender permeates through all our work as rural geographers.

Bibliography

Adler, S. and Brenner, J. (1992), 'Gender and space: lesbians and gay men in the city', *International Journal of Urban and Regional Research*, 16: 24–34.

Agg, J. and Phillips, M. (1998), 'Neglected gender dimensions of rural social restructuring', in Boyle, P. and Halfacree, K. (eds), *Migration into Rural Areas: Theories and Issues*, London: Wiley.

Agyeman, J. and Spooner, R. (1997), 'Ethnicity and the rural environment', in Cloke, P. and Little, J. (eds), *Contested Countryside Cultures: otherness, marginalisation and rurality*, London: Routledge.

Allanson, P. and Whitby, M. (eds) (1996), *The Rural Economy and the British Countryside*, London: Earthscan.

Almas, R. and Haugen, M. (1991), 'Norwegian gender roles in transition: the masculinisation hypothesis in the past and in the future', *Journal of Rural Studies*, 7: 79–84.

Amin, A. (ed.) (1995), *Post-Fordism: A Reader*, Blackwell: Oxford.

Anderson, K. (1997), 'A walk on the wildside', *Progress in Human Geography*, 21: 463–85.

Anderson, K. and Gale, F. (eds) (1992), *Inventing Places: Studies in Cultural Geography*, Melbourne: Longman.

Arkleton Trust (1992), 'Farm Household Adjustment in Western Europe: 1987–1991', Final report on the research programme on Farm Structures and Pluriactivity for the Commission of the European Communities, Brussels.

Aslet, C. (1999), 'The farmer's wife: a twenty-first century economic model?', *Country Life*, 24 June 1999.

Barber, (Lord) (2000), 'The elevation of another three leading ladies', *Country Illustrated*, September 2000: 7–8.

Barrett, M. and Phillips, A. (1992), *Destabilising Theory: Contemporary Feminist Debates*, Cambridge: Polity Press.

Bell, C. and Newby, H. (1971), *Community Studies: an introduction to the sociology of the local community*, London: Allen and Unwin.

Bell, D. (2000a), 'Farm boys and wild men: rurality, masculinity and homosexuality', *Rural Sociology*, 65: 547–61.

Bell, D. (2000b), 'Eroticizing the rural', in Phillips, R., Watt, D. and Shuttleton, D. (eds), *De-centring Sexualities: Politics and Representations Beyond the Metropolis*, London: Routledge.

Bell, D. and Valentine, G. (1995a), *Mapping Desire: Geographies of Sexualities*, London: Routledge.

Bell, D. and Valentine, G. (1995b), 'Queer country: rural lesbian and gay lives', *Journal of Rural Studies*, 11: 113–22.

Bell, D., Binnie, J., Cream, J. and Valentine, G. (1994), 'All hyped up and no place to go', *Gender, Place and Culture*, 1: 31–47.

Bell, M. (1994), *Childerley: Nature and Morality in a Country Village*, Chicago: University of Chicago Press.

Bennett, K. (2000), 'Inter/viewing and inter/subjectivities: powerful performances', in Hughes, A., Morris, C. and Seymour, S. (eds), *Ethnography and Rural Research*, Cheltenham: The Countryside and Community Press.

Bennett, K., Hudson, R. and Benyon, H. (2000), 'Coalfields Regeneration: Dealing with the Consequences of Industrial Decline', Report to the Joseph Rowntree Foundation, York: JRF.

Bialeschki, M.D. and Hicks, H. (1998), ' "I refuse to live in fear": the influence of fear of violence on women's outdoor recreation activities', annual conference of the Leisure Studies Association, Leeds Metropolitan University.

Biehl, J. (1991), *Rethinking Ecofeminist Politics*, Boston, MA: South End Press.

Binnie, J. (1995), 'Trading places: consumption, sexuality and the production of queer space', in Bell, D. and Valentine, G. (eds), *Mapping Desire: Geographies of Sexuality*, London: Routledge.

Binnie, J. (2000), 'Cosmopolitanism and the sexed city', in Bell, D. and Haddour, A. (eds), *City Visions*, Harlow: Pearson.

Binnie, J. and Valentine, G. (1999), 'Geographies of sexuality – a review of progress', *Progress in Human Geography*, 175–87.

Bly, R. (1990), *Iron John: A Book About Men*, Reading, MA: Addison-Wesley.

Boddy, M. and Fudge, C. (1984) (eds), *Local Socialism?*, London: Routledge.

Bondi, L. (1993), 'Locating identity politics', in Keith, M. and Pile, S. (eds), *Place and the Politics of Identity*, London: Routledge.

Bonnett, A. (1996), 'The New Primitives: identity, landscape and cultural appropriation in the mythopoetic men's movement', *Antipode*, 28: 273–91.

Bordo, S. (1990), 'Feminism, postmodernism and gender scepticism', in Nicholson, L. (ed.), *Feminism/Postmodernism*, London: Routledge.

Bouquet, M. (1987), 'Bed, breakfast and evening meal: commensality in the nineteenth and twentieth century farm household in Hartland', in Bouquet, M. and Winter, M. (eds), *Who From Their Labours Rest? Conflict and Practice in Rural Tourism*, Aldershot: Avebury.

Bowlby, S. (1988), 'From corner shop to hypermarket: women and food retailing', in Little, J., Peake, L. and Richardson, P. (eds), *Women in Cities: Gender and the Urban Environment*, London: Macmillan.

Bowlby, S. (1992), 'Feminist geography and the changing curriculum', *Geography*, 77: 349–60.

Bowlby, S., Foord, J. and McDowell, L. (1986), 'The place of locality in gender studies', *Area*, 18: 327–31.

Bowlby, S., Lewis, J., McDowell, L. and Foord, J. (1989), 'The geography of gender', in Peet, R. and Thrift, N. (eds), *New Models in Geography 2*, London: Unwin Hyman.

Boyle, P., Halfacree, K. and Robinson, V. (1998), *Exploring Contemporary Migration*, London: Longman.

Brace, C. (1995), 'Cotswold – that great king of shepherds: masculinist representations of the Cotswolds over three centuries', paper presented at the Women/Time/Space conference, University of Lancaster, March 1995.

Brace, C. (1998), ' "The Legacy of England": dust jacket art and the formation of English national identity c 1920–1950', paper presented at the Association of Art Historians Annual Conference, University of Exeter, April 1998.

Brace, C. (1999), 'Finding England everywhere: regional identity and the construction of national identity, 1890–1940', *Ecumene*, 6: 90–109.

Bradley, T. (1986), 'Poverty and dependency in village England', in Lowe, P., Bradley, T. and Wright, S. (eds), *Deprivation and Welfare in Rural Areas*, Norwich: Geo Books.

Bradley, T. and Lowe, P. (eds) (1984), *Locality and Rurality: Economy and Society in Rural Regions*, Norwich: Geo Books.

Braithwaite, M. (1994), 'The Economic Role and Situation of Women in Rural Areas', Report to the European Commission, Brussels.

Brandth, B. (1994), 'Changing femininity: the social construction of women farmers in Norway', *Sociologia Ruralis*, 34: 127–49.

Brandth, B. (1995), 'Rural masculinity in transition: gender images in tractor advertisements', *Journal of Rural Studies*, 11: 123–33.

Brandth, B. and Bolsø, A. (1994), 'Men, women and biotechnology: a feminist "care" ethic in agricultural science?', in Whatmore, S., Marsden, T. and Lowe, P. (eds), *Gender and Rurality*, London: David Fulton.

Brandth, B. and Haugen, M. (1998), 'Breaking into a masculine discourse. Women and farm forestry', *Sociologia Ruralis*, 38: 427–42.

Brandth, B. and Haugen, M. (2000), 'From lumberjack to business manager: masculinity in the Norwegian forestry press', *Journal of Rural Studies*, 16: 343–56.

Brownill, S. and Halford, S. (1990), 'Understanding women's involvement in local politics', *Political Geography Quarterly*, 9: 396–414.

Bryant, L. (1999), 'Body politics at work: exploring gender and sexuality in constructions of occupation(s) in agriculture', paper presented to the Gender and Transformations in Rural Europe conference, University of Wageningen, The Netherlands, October 1999.

Buckingham-Hatfield, S. and Percy, S. (eds) (1998), *Constructing Local Environmental Agendas*, London: Routledge.

Burgess, J. (1995), 'Growing in confidence: understanding people's perceptions of urban fringe woodlands', Report to the Countryside Commission, Cheltenham: Countryside Commission.

Butler, J. (1990), *Gender Trouble: Feminism and the Subversion of Identity*, London: Routledge.

Butler, J. (1993), *Bodies that Matter: on the Discursive Limits of 'Sex'*, London: Routledge.

Butler, R. (1999), 'The body', in Cloke, P., Crang, P. and Goodwin, M. (eds), *Introducing Human Geographies*, London: Arnold.

Butler, R. and Bowlby, S. (1997), 'Bodies and spaces: an exploration of disabled people's use of space', *Environment and Planning D: Society and Space*, 15: 411–33.

Castells, M. (1983), *The City and the Grassroots*, Berkeley: University of California Press.

Castree, N. (1995), 'The nature of produced nature', *Antipode*, 27: 12–48.

Cavanagh, J. (1999), 'Others and structures in the post(-)rural: degrees of separation', unpublished PhD thesis, Department of Geography, University of Exeter.

Champion, T. and Watkins, C. (1991), *People in the Countryside: Studies of Social Change in Rural Britain*, London: Paul Chapman.

Chapman, P., Phimister, E., Shucksmith, M., Upward, R. and Vera-Toscano, E. (1998), 'Poverty and Exclusion in Rural Britain: The Dynamics of Low Income and Employment', Report to the Joseph Rowntree Foundation, York: JRF.

Cheney, J. (1985), *Lesbian Land*, Minneapolis, MN: Word Weavers.

Cloke, P. (1989), 'Rural geography and political economy', in Peet, R. and Thrift, N. (eds), *New Models in Geography: The Political Economy Perspective*, London: Unwin Hyman.

Cloke, P. (1994), '(En)culturing political economy: a day in the life of a "rural geographer"', in Cloke, P., Doel, M., Matless, D., Phillips, M. and Thrift, N. (eds), *Writing the Rural: Five Cultural Geographies*, London: Paul Chapman.

Cloke, P. (1997), 'Country backwater to virtual village? Rural studies and "The Cultural Turn"', *Journal of Rural Studies*, 13: 367–75.

Cloke, P. and Goodwin, M. (1992), 'Conceptualising countryside change: from post-Fordism to rural structured coherence', *Transactions of the Institute of British Geographers*, 17: 321–36.

Cloke, P. and Little, J. (1990), *The Rural State? Limits to Planning in Rural Society*, Oxford: Oxford University Press.

Cloke, P. and Little, J. (eds) (1997), *Contested Countryside Cultures: Otherness, Marginality and Rurality*, London: Routledge.

Cloke, P. and Perkins, H. (1998), 'Representations of adventure tourism in New Zealand', *Environment and Planning D: Society and Space*, 16: 185–218.

Cloke, P. and Thrift, N. (1990), 'Class change and conflict in rural areas', in Marsden, T., Lowe, P. and Whatmore, S. (eds), *Rural Restructuring*, London: David Fulton.

Cloke, P., Milbourne, P. and Thomas, C. (1994), 'Lifestyles in Rural England', Rural Development Commission Research Report 18, Salisbury: RDC.

Cloke, P., Phillips, M. and Thrift, N. (1995), 'The new middle classes and the social constructs of rural living', in Butler, T. and Savage, M. (eds), *Social Change and the Middle Classes*, London: UCL Press.

Cook, I. and Crang, M. (1995), *Doing Ethnographies*, CATMOG Series.

Cook, I., Crang, P. and Thorpe, M. (1999), 'Eating into Britishness: multicultural imaginaries and the identity politics of food', in Roseneil, S. and Seymour, J. (eds), *Practicing Identities: Power and Resistance*, London: Macmillan.

Cook, I., Crouch, D., Naylor, S. and Ryan, J. (eds) (2000), *Cultural Turns/Geographical Turns: perspectives on cultural geography*, Harlow: Pearson.

Coole, D. (1997), 'Feminism without nostalgia', *Radical Philosophy*, 83: 17–24.

Coote, A. and Pattullo, P. (1990), *Power and Prejudice: women and politics*, London: Weidenfeld and Nicolson.

Cosgrove, D. (1985), 'Prospect, perspective and the evolution of the landscape idea', *Transactions of the Institute of British Geographers*, 10: 45–62.

Cosgrove, D. and Daniels, S. (eds) (1988), *The Iconography of Landscape: essays on the symbolic, design and use of past environments*, Cambridge: Cambridge University Press.

Crang, P., Dwyer, C., Prinjha, S. and Jackson, P. (2000), 'Geographies of transnationalism: negotiating nation and nationalism at a distance', paper presented at the annual conference of the Royal Geographical Society/Institute of British Geographers, University of Sussex, Brighton, UK, January 2000.

Cresswell, T. (1996), *In Place/Out of Place: geography, ideology and transgression*, Minneapolis: University of Minnesota Press.

Cronon, W. (ed.) (1995), *Uncommon Ground: Towards Reinventing Nature*, New York: W.W. Norton and Co.

Crouch, D. (1992), 'Popular culture and what we make of the rural, with a case study of village allotments', *Journal of Rural Studies*, 8: 229–40.

D'Augelli, A. and Hart, M. (1987), 'Gay women, men and families in rural settings', *American Journal of Community Psychology*, 15: 79–93.

Dahlström, M. (1996), 'Young women in a male periphery – experiences from the Scandinavian north', *Journal of Rural Studies*, 12: 259–72.

Daniels, S. (1993), *Fields of Vision: landscape imagery and national identity in England and the United States*, Cambridge: Polity Press.

Daniels, S. (1999), *Joseph Wright*, London: The Tate Gallery.

Davidoff, L., L'Esperance, J. and Newby, H. (1976), 'Landscape with figures: home and community in English society', in Mitchell, J. and Oakley, A. (eds), *The Rights and Wrongs of Women*, Harmondsworth: Penguin.

Day, G. and Murdoch, J. (1993), 'Locality and community: coming to terms with place', *Sociological Review*, 41: 82–111.

Delphy, C. (1984), *Close to Home: a Materialist Analysis of Women's Oppression*, London: Hutchinson.

Dempsey, K. (1992), *A Man's Town: Inequality between women and men in rural Australia*, Melbourne: Oxford University Press.

Department of the Environment/Ministry of Agriculture Fisheries and Food (1995), 'Rural England: a Nation Committed to Living in the Countryside', Government White Paper, London: HMSO.

Doel, M. (1994), 'Something resists: reading – deconstruction as ontological infestation (departures from the texts of Jacques Derrida)', in Cloke, P., Doel, M., Matless, D., Phillips, M. and Thrift, N., *Writing the Rural: Five Cultural Geographies*, London: Paul Chapman.

Dowler, L. (1999), 'Participant Observation', in McDowell, L. and Sharp, J. (eds), *A Feminist Glossary of Human Geography*, London: Arnold.

Duncan, S. and Goodwin, M. (1988), *The Local State and Uneven Development*, Cambridge: Polity Press.

Dwyer, C. (1999), 'Negotiations of femininity and identity for young British Muslim women', in Laurie, N., Dwyer, C., Holloway, S. and Smith, F. (eds), *Geographies of New Femininities*, London: Longman.

Edwards, W. and Woods, M. (2000), 'Parish pump politics: participation, power and rural community governance', paper presented at the annual conference of the Royal Geographical Society/Institute of British Geographers, University of Sussex, January 2000.

Errington, A., Bennett, R. and Marshall, B. (1989), 'Employment and Training in Rural Areas', Rural Development Commission Report No. 3, Salisbury: RDC.

Evans, N. and Ilbery, B. (1992), 'Farm-based accommodation and the restructuring of agriculture: evidence from three English counties', *Journal of Rural Studies*, 8: 85–96.

Evans, N. and Ilbery, B. (1996), 'Exploring the influence of farm-based pluriactivity on gender relations in capitalist agriculture', *Sociologia Ruralis*, 36: 74–92.

Evans, R. (1999), ' "You questioning my manhood boy?" Masculine identity, performativity and the performance of work in a rural staples economy', paper presented at the Gender and Rural Transformations in Europe Conference, University of Wageningen, The Netherlands, October 1999.

Faludi, S. (1992), *Backlash, The Undeclared War Against Women*, London: Chatto and Windus.

Fellows, W. (1996), *Farm Boys: lives of gay men from the rural Midwest*, Madison: University of Wisconsin Press.

Fielding, S. (2000), 'The importance of being Shaun: self-reflection and ethnography', in Hughes, A., Morris, C. and Seymour, S. (eds), *Ethnography and Rural Research*, Cheltenham: The Countryside and Community Press.

Fitzsimmonds, M. (1989), 'The matter of nature', *Antipode*, 21: 106–20.

Foord, J. and Gregson, N. (1986), 'Patriarchy: towards a reconceptualisation', *Antipode*, 18: 181–211.

Foord, J. and Lewis, J. (1984), 'New towns and gender relations in old industrial regions: women's employment in Peterlee and East Kilbride', *Built Environment*, 10: 42–52.

Fothergill, S., Gudgin, G., Kitson, M. and Monk, S. (1985), 'Rural industrialisation: trends and cases', in Healey, B. and Ilbery, B. (eds), *The Industrialisation of the Countryside*, Norwich: Geo Books.

Frankenberg, R. (1966), *Communities in Britain*, Harmondsworth: Penguin.

Gans, H. (1962), *The Urban Villagers: group and class in the life of Italian-Americans*, New York: Free Press.

Gardner, B. (1996), *European Agriculture: Policies, Production and Trade*, London: Routledge.

Gasson, R. (1980), 'Roles of farm women in England', *Sociologia Ruralis*, 20: 165–80.

Gasson, R. (1992), 'Farmers' wives and their contribution to the farm business', *Journal of Agricultural Economics*, 43: 74–87.

Gasson, R. and Winter, M. (1992), 'Gender relations and farm household pluriactivity', *Journal of Rural Studies*, 8: 387–97.

Goodman, D. and Redclift, M. (1991), *Refashioning Nature: Food, Ecology and Culture*, London: Routledge.

Goodman, D., Sorj, B. and Wilkinson, J. (1987), *From Farming to Biotechnology: A Theory of Agro-industrial Development*, Oxford: Blackwell.

Goodwin, M. (1998), 'The governance of rural areas: some emerging research issues and agendas', *Journal of Rural Studies*, 14: 5–12.

Goodwin, M. and Painter, J. (1996), 'Local governance, the crisis of Fordism and the changing geographies of regulation', *Transactions of the Institute of British Geographers*, 21: 635–48.

Goss, J. (1996), 'Introduction to focus groups', *Area*, 28: 113–14.

Greer, G. (1999), *The Whole Woman*, London: Tavistock.

Gruffudd, P. (1994), 'Selling the countryside: representations of rural Britain', in Gold, J. and Ward, S. (eds), *Place Promotion: the use of publicity and marketing to sell towns and regions*, Chichester: Wiley.

Haartsen, T., Groote, P. and Huigen, P. (eds) (2000), *Claiming Rural Identities*, Assen, The Netherlands: Van Gorcum.

Halfacree, K. (1993), 'Locality and social representation: space, discourse and the alternative definitions of the rural', *Journal of Rural Studies*, 9: 23–37.

Halfacree, K. (1994), 'The importance of "the rural" in the constitution of counterurbanisation: evidence from England in the 1980s', *Sociologia Ruralis*, 34: 168–89.

Halfacree, K. (1996), 'Out of place in the country: travellers and the "rural idyll"', *Antipode*, 28: 42–71.

Halford, S. (1989), 'Spatial divisions and women's initiatives in British local government', *Geoforum*, 20: 161–74.

Halliday, J. (1997), 'Children's services and care: a rural view', *Geoforum*, 28: 103–19.

Halliday, J. and Little, J. (in press), 'Amongst women: exploring the reality of rural childcare', submitted to *Sociologia Ruralis*.

Hanrahan, P. and Cloke, P. (1983), 'Towards a critical appraisal of rural settlement planning in England and Wales', *Sociologia Ruralis*, 23: 109–29.

Hanson, S. and Pratt, G. (1995), *Gender, Work and Space*, London: Routledge.

Henriques, R. (1950), *The Cotswolds*, London: P. Elek.

Hogbacka, R. (1995), 'Women's work and well-being in different types of rural areas in Finland', paper presented at the XVI Congress of European Society for Rural Sociology, Prague.

Hoggart, K. (1990), 'Lets do away with the rural', *Journal of Rural Studies*, 6: 245–57.

Hoggart, K. (1997), 'The middle classes in rural England 1971–1991', *Journal of Rural Studies*, 13: 253–73.

Hoggart, K., Buller, H. and Black, R. (1995), *Rural Europe: identity and change*, London: Arnold.

Houlton, D. and Short, B. (1995), 'Sylvanian Families: the production and consumption of the rural community', *Journal of Rural Studies*, 11: 367–85.

Hughes, A. (1996), 'Women in the countryside: gender identities and rurality', unpublished PhD thesis, University of Bristol.

Hughes, A. (1997a), 'Women and rurality: gendered experiences of "community" in village life', in Milbourne, P. (ed.), *Revealing Rural 'Others': Representation, Power and Identity in the British Countryside*, London: Pinter.

Hughes, A. (1997b), 'Rurality and cultures of womanhood: domestic identities and moral order in rural life', in Cloke, P. and Little, J. (eds), *Contested Countryside Cultures: otherness, marginalisation and rurality*, London: Routledge.

Hughes, A., Morris, C. and Seymour, S. (eds) (2000), *Ethnography and Rural Research*, Cheltenham: The Countryside and Community Press.

Imrie, R. (1996), 'Ableist geographers, disablist spaces: towards a reconstruction of Golledge's "Geography and the disabled" ', *Transactions of the Institute of British Geographers New Series*, 21: 397–403.

Jackson, P. (1983), 'Principles and problems of participant observation', *Geografiska Annaler B*, 65: 39–46.

Jackson, P. (1999), 'Identity', in McDowell, L. and Sharp, J. (eds), *A Feminist Glossary of Human Geography*, London: Arnold.

Jessop, B. (1993), 'Towards a Schumpeterian workfare state? Preliminary remarks on post-Fordist political economy', *Studies in Political Economy*, 40: 7–39.

Jessop, B. (1995), 'The regulation approach, governance and post-Fordism: alternative perspectives on economic and political change', *Economy and Society*, 24: 307–33.

Johnston, L. (1996), 'Flexing femininity: female body-builders refiguring "the body" ', *Gender, Place and Culture*, 3: 327–40.

Jones, O. (1995), 'Lay discourses of the rural: developments and implications for rural studies', *Journal of Rural Studies*, 11: 35–49.

Jones, O. (1997), 'Little figures, big shadows: country childhood stories', in Cloke, P. and Little, J. (eds), *Contested Countryside Cultures: Otherness, Marginalisation and Rurality*, London: Routledge.

Jones, O. (1999), 'Tomboy tales: the rural, gender and the nature of childhood', *Gender, Place and Culture*, 6: 117–36.

Jones, O. (2000), 'Is all well in Allswell? (a cute, exclusive village in SW England)', paper presented at the annual conference of the Rural Economy and Society Study Group, University of Exeter, September 2000.

Jones, O. and Little, J. (2000), 'Rural Challenge(s): partnership and new rural governance', *Journal of Rural Studies*, 16: 171–84.

Keeble, D. (1984), 'The urban–rural manufacturing shift', *Geography*, 69: 163–6.

Keeble, D. and Tyler, P. (1995), 'Enterprising behaviour and the urban–rural shift', *Urban Studies*, 32: 975–98.

Keith, M. and Pile, S. (1993), *Place and the Politics of Identity*. London: Routledge.

Knopp, L. (1990), 'Some theoretical implications of gay involvement in an urban land market', *Political Geography Quarterly*, 9: 337–52.

Kolodny, A. (1975), *The Lay of the Land: Metaphor as experience in American Life and Letters*, Chapel Hill: University of North Carolina Press.

Kramer, J. (1995), 'Bachelor farmers and spinsters: gay and lesbian identities and communities in rural North Dakota', in Bell, D. and Valentine, G. (eds), *Mapping Desire: geographies of sexualities*, London: Routledge.

Laing, S. (1992), 'Images of the rural in popular culture', in Short, B. (ed.), *The Rural Community: Image and Analysis*, Cambridge: Cambridge University Press.

Larsen, S.E. (1994), 'Nature on the move: meanings of nature in contemporary culture', *Ecumene*, 1: 283–300.

Lauria, M. (ed.) (1997), *Reconstructing Urban Regime Theory: regulating urban politics in a global economy*, Thousand Oaks, CA: Sage.

Lauria, M. and Knopp, L. (1985), 'Towards an analysis of gay communities in the urban renaissance', *Urban Geography*, 6: 152–69.

Laurie, N., Dwyer, C., Holloway, S. and Smith, F. (eds) (1999), *Geographies of New Femininities*, London: Longman.

Lawrence, M. (1998), 'Miles from home in the field of dreams: rurality and the social at the end of history', *Environment and Planning D. Society and Space*, 16: 705–32.

Leach, B. (1999), 'Transforming rural livelihoods: gender, work and restructuring in three Ontario communities', in Neysmith, S. (ed.), *Restructuring Caring Labour: Discourse, State Practice and Everyday Life*, New York: Oxford University Press.

Leidner, R. (1991), 'Selling hamburgers and selling insurance', *Gender and Society*, 5: 154–77.

Leonard, M. (1998), 'Paper planes: travelling the new grrl geographies', in Skelton, T. and Valentine, G. (eds), *Cool Places. Geographies of Youth Culture*, London: Routledge.

Leyshon, M. (2000), 'The betweeness of being a rural youth: inclusive and exclusive lifestyles', paper presented at the annual conference of the Rural Economy
and Society Study Group, University of Exeter, September 2000.

Leyshon, M. (forthcoming), 'Youth identity, culture and marginalisation in the
countryside', PhD thesis, Department of Geography, University of Exeter.

Lipiens, R. (1998), 'Women of broad vision: nature and gender in the environmental activism of Australia's "Women in Agriculture" movement', *Environment and Planning A*, 30: 1179–96.

Lipiens, R. (2000a), 'New energies for old: reworking approaches to "community" in contemporary rural studies', *Journal of Rural Studies*, 16: 23–36.

Lipiens, R. (2000b), 'Exploring rurality through "community": discourses, practices and spaces shaping Australian and New Zealand rural "communities"',
Journal of Rural Studies, 16: 325–42.

Little, J. (1986), 'Feminist perspectives in rural geography: an introduction',
Journal of Rural Studies, 2: 1–8.

Little, J. (1987), 'Gender relations in rural areas: the importance of women's
domestic role', *Journal of Rural Studies*, 3, 335–42.

Little, J. (1991), 'Women in the rural labour market: a policy evaluation', in
Champion, T. and Watkins, C. (1991), *People in the Countryside: Studies of
Social Change in Rural Britain*, London: Paul Chapman.

Little, J. (1994a), 'Women's initiatives in town planning in England', *Town
Planning Review*, 65: 261–76.

Little, J. (1994b), *Gender, Planning and the Policy Process*, London: Pergamon.

Little, J. (1997a), 'Employment, marginality and women's self-identity', in Cloke, P.
and Little, J. (eds), *Contested Countryside Cultures: Otherness, Marginalisation
and Rurality*, London: Routledge.

Little, J. (1997b), 'Women and voluntary work in rural communities', *Gender,
Place and Culture*, 4: 197–209.

Little, J. (1999), 'Otherness, representation and the cultural construction of
rurality', *Progress in Human Geography*, 23: 437–42.

Little, J. and Austin, P. (1996), 'Women and the rural idyll', *Journal of Rural
Studies*, 12: 101–11.

Little, J. and Jones, O. (2000), 'Masculinities, gender and rural policy', *Rural
Sociology*, 65: 621–39.

Little, J. and Jones, O. (forthcoming), 'Competing for rural regeneration'.

Little, J. and Leyshon, M. (1998), *A Place to Hang Out: rural youth and access
to education, training and leisure in Somerset*, Report to the Somerset Rural
Youth Project, Taunton: Somerset County Council.

Little, J. and Pollard, J. (2000), 'Making scents of the body economic: perfume,
consumption and gender identity', paper presented at the annual conference of
the Royal Geographical Society/Institute of British Geographers, University of
Sussex, Brighton, UK, January 2000.

Little, J., Ross, K. and Collins, I. (1991), *Women and Employment in Rural
Areas*, Rural Development Commission Research Report 10, Salisbury: RDC.

Little, J., Clemments, J. and Jones, O. (1998), 'Rural Challenge and the Changing Culture of Rural Regeneration Policy', in Oatley, N. (ed.), *Cities, Economic Competition and Urban Policy*, London: Paul Chapman.

Littlejohn, J. (1963), *Westrigg. The Sociology of a Cheviot Parish*, London: Routledge and Kegan Paul.

Longhurst, R. (1997), '(Dis)embodied geographies', *Progress in Human Geography*, 21: 486–501.

Longhurst, R. (2000), ' "Corporeographies" of pregnancy: "bikini babes" ', *Environment and Planning D: Society and Space*, 18: 453–72.

Lowe, P., Cox, G., MacEwan, M., O'Riordan, T. and Winter, M. (1986), *Countryside Conflicts: the politics of farming, forestry and conservation*, Aldershot: Gower.

Lowe, P., Clark, J., Seymour, S. and Ward, N. (1997), *Moralizing the Environment: Countryside Change, Farming and Pollution*, London: UCL Press.

McDowell, L. (1983), 'Towards an understanding of the gender division of urban space', *Environment and Planning D: Society and Space*, 1: 15–30.

McDowell, L. (1986), 'Beyond patriarchy: a class-based examination of women's subordination', *Antipode*, 18: 311–21.

McDowell, L. (1991), 'The baby and the bath water: deconstruction, diversity and feminist theory in geography', *Geoforum*, 22: 123–34.

McDowell, L. (1992), 'Doing gender: feminism, feminists and research methods in human geography', *Transactions of the Institute of British Geographers*, 17: 399–418.

McDowell, L. (1999), *Gender, Identity and Place: Understanding Feminist Geographies*, Cambridge: Polity Press.

McDowell, L. and Court, G. (1994), 'Missing subjects: gender, sexuality and power in merchant banking', *Economic Geography*, 70: 225–51.

McDowell, L. and Peake, L. (1990), 'Women in geography revisited', *Journal of Geography in Higher Education*, 14: 19–30.

McDowell, L. and Sharp, J. (eds) (1997), *Space, Gender, Knowledge: Readings in Feminist Geography*, London: Arnold.

MacFayden, D. and Hole, C. (1983), *Folk Customs of Britain*, London: Hutchinson.

McKenzie, S. and Rose, D. (1983), 'Industrial change, the domestic economy and home life', in Anderson, J., Duncan, S. and Hudson, R. (eds), *Redundant Spaces? Social Change and Industrial Decline in Cities and Regions*, London: Academic Press.

McLafferty, S. (1995), 'Counting for women', *Professional Geographer*, 47: 436–42.

McLaughlin, B. (1989), 'The rhetoric and reality of rural deprivation', *Journal of Rural Studies*, 2: 291–307.

MacNaughten, P. and Urry, J. (1998), *Contested Natures*, London: Sage.

Mancoske, R. (1997), 'Rural HIV/AIDS social services for gays and lesbians', *Journal of Gay and Lesbian Social Services*, 7: 37–52.

Mansfield, L. and Maguire, J. (1999), 'Active women, power relations and gendered identities: embodied experiences of aerobics', in Roseneil, S. and Seymour, J. (eds), *Practicing Identities: Power and Resistance*, London: Macmillan.

Marsden, T. (1995), 'Beyond agriculture? Regulating the new rural spaces', *Journal of Rural Studies*, 11: 285–96.

Marsden, T. (1998), 'Economic perspectives', in Ilbery, B. (ed.), *The Geography of Rural Change*, Harlow: Addison, Wesley, Longman.

Marsden, T. and Little, J. (eds) (1990), *Political, Social and Economic Perspectives on the International Food System*, Aldershot: Avebury.

Marsden, T. and Murdoch, J. (1998), 'The shifting nature of rural governance and community participation', *Journal of Rural Studies*, 14: 1–4.

Marsden, T., Munton, R., Whatmore, S. and Little, J. (1986), 'Towards a political economy of capitalist agriculture: a British perspective', *International Journal of Urban and Regional Research*, 10: 498–521.

Marsden, T., Murdoch, J., Lowe, P., Munton, R. and Flynn, A. (1993), *Constructing the Countryside*, London: UCL Press.

Marshall, J. (1999), 'Tradition and modernity: changing social and space relations on Grand Manan Island, N.B.', paper presented at the International Rural Geography Symposium, St Mary's University, Halifax, July 1999.

Massey, D. (1994), *Space, Place and Gender*, Cambridge: Polity Press.

Massey, D. and Allen, J. (eds) (1984), *Geography Matters!*, Cambridge: Open University Press.

Matless, D. (1995), 'Doing the English village, 1945–90: An essay in imaginative geography', in Cloke, P., Doel, M., Matless, D., Phillips, M. and Thrift, N. (eds), *Writing the Rural: Five Cultural Geographies*, London: Paul Chapman.

Matless, D. (1998), *Landscape and Englishness*, London: Rektion.

Meinig, D. (ed.) (1979), *The Interpretation of Ordinary Landscapes*, Oxford: Oxford University Press.

Middleham Key Partnership (1994), 'A United Community: the Middleham Key Partnership's Bid for Rural Challenge Funding', Middleham Town Council.

Middleton, A. (1986), 'Marking boundaries: men's space and women's space in a Yorkshire village', in Bradley, T., Lowe, P. and Wright, S. (eds), *Deprivation and Welfare in Rural Areas*, Norwich: Geo Books.

Milbourne, P. (ed.) (1997), *Revealing Rural Others: Representation, Power and Identity in the British Countryside*, New York: Pinter.

Mitchell, D. (2000), *Cultural Geography: A Critical Introduction*, Oxford: Blackwell.

Monk, J. (1992), 'Gender in the landscape: expressions of power and meaning', in Anderson, K. and Gayle, F. (eds), *Inventing Places: studies in cultural geography*, Melbourne: Wiley.

Monk, J. and Hanson, S. (1982), 'On not excluding the other half of the human in geography', *The Professional Geographer*, 34: 11–23.

Morris, C. and Evans, N. (forthcoming), ' "Cheesemakers are always women": gendered representations of farm life in the agricultural press', *Gender, Place and Culture*.

Mort, F. (1989), 'The politics of consumption', in Hall, S. and Jacques, M. (eds), *New Times*, London: Lawrence and Wishart.

Moseley, M. (ed.) (1982), *Power, Planning and People in Rural East Anglia*, Norwich: University of East Anglia.

Murdoch, J. and Abram, S. (1998), 'Defining the limits of community governance', *Journal of Rural Studies*, 14: 41–50.

Murdoch, J. and Marsden, T. (1994), *Reconstituting Rurality: Class, Community and Power in the Development Process*, London: UCL Press.

Murdoch, J. and Pratt, A. (1993), 'Modernism, postmodernism and the post-rural', *Journal of Rural Studies*, 9: 411–27.

Murdoch, J. and Pratt, A. (1997), 'From the power of topography to the topography of power: a discourse on strange ruralities', in Cloke, P. and Little, J. (eds), *Contested Countryside Cultures: Otherness, Marginalisation and Rurality*, London: Routledge.

Nash, C. (1994), 'Remapping the body/land: new cartographies of identity by Irish women artists', in Blunt, A. and Rose, G. (eds), *Writing Women and Space*, London: Guilford Press.

Nash, C. (1996), 'Reclaiming vision: looking at landscape and the body', *Gender, Place and Culture*, 3: 149–70.

Newby, H. (1979), *Green and pleasant land? Social Change in Rural England*, Harmondsworth: Penguin.

Newby, H., Bell, C., Rose, D. and Saunders, P. (1978), *Property, Paternalism and Power*, London: Hutchinson.

Oakley, A. (1981), 'Interviewing women: a contradiction in terms', in Roberts, H. (ed.), *Doing Feminist Research*, London: Routledge and Kegan Paul.

Oatley, N. (ed.) (1998), *Cities, Economic Competition and Urban Policy*, London: Paul Chapman.

Oatley, N. and Lambert, C. (1995), 'Evaluating competitive urban policy: the City Challenge initiative', in Hambleton, R. and Thomas, H. (eds), *Urban Policy Evaluation: Challenge and Change*, London: Paul Chapman.

Oswell, D. (2000), 'Suburban tales, television, masculinity and textual geographies', in Bell, D. and Haddour, A. (eds), *City Visions*, Harlow: Pearson.

Pahl, R. (1965), 'Class and community in an English village', *Sociologia Ruralis*, 6: 299–329.

Painter, J. (1998), 'Entrepreneurs are made, not born: learning and urban regimes in the production of entrepreneurial cities', in Hall, T. and Hubbard, P. (eds), *The Entrepreneurial City: Geographies of Politics, Regime and Regulation*, Wiley: London.

Peck, J. and Tickell, A. (1994), 'Too many partnerships. The future for regeneration partnerships', *Local Economy*, 9: 251–65.

Phillips, M. (1993), 'Rural gentrification and the process of class colonisation', *Journal of Rural Studies*, 9: 123–40.

Phillips, M. (1998a), 'Social perspectives', in Ilbery, B. (ed.), *The Geography of Rural Change*, Harlow: Addison, Wesley, Longman.

Phillips, M. (1998b), 'The restructuring of social imaginations in rural geography', *Journal of Rural Studies*, 14: 121–53.

Phillips, M. (2000), 'Theories of positionality and ethnography in the rural', in Hughes, A., Morris, C. and Seymour, S. (eds), *Ethnography and Rural Research*, Cheltenham: The Countryside and Community Press.

Phillips, R. (1995), 'Spaces of adventure and cultural politics of masculinity: R.M. Ballantyne and the *Young Fur Traders*', *Environment and Planning D: Society and Space*, 13: 591–608.

Philo, C. (1992), 'Neglected rural geographies: a review', *Journal of Rural Studies*, 8: 193–207.

Philo, C. (1997), 'Of other rurals?', in Cloke, P. and Little, J. (eds), *Contested Countryside Cultures: Otherness, Marginalisation and Rurality*, London: Routledge.

Pickup, L. (1988), 'Hard to get around: a study of women's travel mobility', in Little, J., Peake, L. and Richardson, P. (eds), *Women in Cities: Gender and the Urban Environment*, London: Macmillan.

Pratt, G. (1999), 'Geographies of identity and difference: marking boundaries', in Massey, D., Allen, J. and Sarre, P. (eds), *Human Geography Today*, Cambridge: Polity Press.

Pringle, R. (1999), 'Sexuality', in McDowell, L. and Sharp, J. (eds), *A Feminist Glossary of Human Geography*, London: Arnold.

Purvis, L. (1996), *A Long Walk in Wintertime*, London: Hodder and Stoughton.

Pye-Smith, C. and Rose, C. (1984), *Crisis and Conservation: Conflict in the British Countryside*, Harmondsworth: Penguin.

Redclift, N. (1985), 'The contested domain: gender, accumulation and the labour process', in Redclift, N. and Mingione, E. (eds), *Beyond Employment: household, gender and subsistence*, Oxford: Blackwell.

Redclift, N. and Whatmore, S. (1990), 'Household consumption and livelihood: ideologies and issues in rural research', in Marsden, T., Lowe, P. and Whatmore, S. (eds), *Rural Restructuring, Global Processes and their Responses*, London: David Fulton.

Revill, G. and Seymour, S. (2000), 'Telling stories: story telling as a textual strategy', in Hughes, A., Morris, C. and Seymour, S. (eds), *Ethnography and Rural Research*, Cheltenham: The Countryside and Community Press.

Rich, A. (1986), *Blood, Bread and Poetry: selected prose 1979–1985*, London: Norton.

Riley, K. (1990), 'Equality for women: the role of local authorities', *Local Government Studies*, Jan/Feb: 49–68.

Rivers, M.-J. (1992), 'The Contribution of Women to the Rural Economy', Ministry of Agriculture and Fisheries, Wellington, New Zealand.

Roberts, H. (ed.) (1981), *Doing Feminist Research*, London: Routledge and Kegan Paul.

Rogers, A. (1987), 'Voluntarism, self-help and rural community development: some current approaches', *Journal of Rural Studies*, 3: 353–60.

Rose, G. (1993), *Feminism and Geography: The Limits of Geographical Knowledge*, Cambridge: Polity Press.

Rose, G. (1996), 'Geography as a science of observation: the landscape, the gaze and masculinity', in Agnew, J., Livingstone, D. and Rogers, A. (eds), *Human Geography: an essential anthology*, Oxford: Blackwell.

Roseneil, S. and Seymour, J. (eds) (1999), *Practicing Identities: Power and Resistance*, London: Macmillan.

Rural Development Commission (1998), *Rural Challenge: Lessons for the Future*, Salisbury: RDC.

Rural Sociology (2000), 'Special Issue', December: 64.

Sachs, C. (1983), *Invisible Farmers: Women's Work in Agricultural Production*, Totowa, NJ: Rhinehart Allenheld.

Sachs, C. (1991), 'Women's work and food: a comparative perspective', *Journal of Rural Studies*, 7: 49–56.

Sachs, C. (1994), 'Rural women's environmental activism in the USA', in Whatmore, S., Marsden, T. and Lowe, P. (eds), *Gender and Rurality*, London: David Fulton.

Saugeres, L. (1998), 'Representations of femininity and masculinity: gender relations and identities amongst farm families in a French community', unpublished PhD thesis, Manchester Metropolitan University.

Saugeres, L. (forthcoming), 'The cultural representation of the farming landscape: masculinity, power and nature', *Gender, Place and Culture*.

Segal, L. (1999), *Why Feminism?*, Cambridge: Polity Press.

Seymour, S. and Short, C. (1994), 'Gender, church and people in rural areas', *Area*, 26: 45–56.

Short, J. (1991), *Imagined Country: Society, Culture and Environment*, London: Routledge.

Shortall, S. (1992), 'Power analysis and farm wives: an empirical study of the power relations affecting women on Irish farms', *Sociologia Ruralis*, XXXII: 431–51.

Shuttleton, D. (2000), 'The queer politics of gay pastoral', in Phillips, R., Watt, D. and Shuttleton, D. (eds), *De-centring sexualities: politics of representation beyond the metropolis*, London: Routledge.

Silverstein, C. (1981), *Man to Man: gay couples in America*, New York: Quill.

Skeggs, B. (1994), 'Situating the production of feminist ethnography', in Maynard, M. and Purvis, J., *Researching Women's Lives*, Basingstoke: Taylor and Francis.

Stebbing, S. (1984), 'Women's roles and rural society', in Bradley, T. and Lowe, P. (eds), *Locality and Rurality: Economy and Society in Rural Regions*, Norwich: Geo Books.

Stone, M. (1990), *Rural Childcare*, Rural Development Commission Research Report 9, Salisbury: RDC.

Symes, D. (1991), 'Changing gender roles in productionist and post-productionist agriculture', *Journal of Rural Studies*, 7: 85–90.

Symes, D. and Marsden, T. (1983), 'Complementary roles and asymmetrical lives: farmers' wives in a large farm environment', *Sociologia Ruralis*, 23: 229–41.

Teather, E. (1992), 'The first women's network in NSW: seventy years of the Country Women's Association', *Australian Geographer*, 23: 164–76.

Teather, E. (1994), 'Contesting rurality: country women's social and political networks', in Whatmore, S., Marsden, T. and Lowe, P. (eds), *Gender and Rurality*, Critical Perspectives on Rural Change Series V1, London: Fulton.

The Rural Group of Labour MPs (1999), *Rural Audit: health check on rural Britain*, London: House of Commons.

Tickell, A. and Peck, J. (1996), 'Return of the Manchester Men: men's words and men's deeds in the remaking of the local state', *Transactions of the Institute of British Geographers*, 21: 595–616.

Till, K. (1999), 'Landscape', in McDowell, L. and Sharp, J. (eds), *A Feminist Glossary of Human Geography*, London: Arnold.

Tivers, J. (1985), *Women Attached: The Daily Lives of Women With Young Children*, London: Croom Helm.

Townsend, A. (1991), 'New forms of employment in rural areas: a national perspective', in Champion, T. and Watkins, C. (eds), *People in the Countryside*, London: Paul Chapman.

Trollop, J. (1989), *A Village Affair*, London: Bloomsbury.

Troughton, M. (1995), 'Rural Canada and Canadian rural geography: an appraisal', *The Canadian Geographer*, 39: 290–305.

Urry, J. (1984), 'Capitalist restructuring, competition and the regions', in Bradley, T. and Lowe, P. (eds), *Locality and Rurality: Economy and Society in Rural Regions*, Norwich: Geo Books.

Valentine, G. (1992), 'Coping with fear of male violence: women's use of precautionary behaviour in public space', paper presented at the Anglo-German conference on Women and the City, Hamburg, April 1992.

Valentine, G. (1993), '(Hetero)sexing space: lesbian perceptions and experiences of everyday space', *Environment and Planning D: Society and Space*, 11: 395–413.

Valentine, G. (1997), 'Making space: lesbian separatist communities in the United States', in Cloke, P. and Little, J. (eds), *Contested Countryside Cultures: Otherness, Marginality and Rurality*, London: Routledge.

Valentine, G. (1999), 'A corporeal geography of consumption', *Environment and Planning D: Society and Space*, 17: 329–51.

Walby, S. (1986), *Patriarchy at Work*, Cambridge: Polity Press.

Walby, S. (ed.) (1988), *Gender Segregation at Work*, Milton Keynes: Open University Press.

Walby, S. (1990), *Theorizing Patriarchy*, Oxford: Blackwell.

Walby, S. (1997), *Gender Transformations*, London: Routledge.

Ward, N. and McNicholas, K. (1998), 'Reconfiguring rural development in the UK: objective 5b and the new rural governance', *Journal of Rural Studies*, 14: 27–40.

Watkins, F. (1998), 'Cultural constructions of rurality', unpublished PhD thesis, University of Sheffield.

Watson, S. (ed.) (1990), *Playing the State: Australian feminist interventions*, Sydney: Allen and Unwin.

Weston, K. (1995), 'Get thee to a big city: sexual imaginary and the great gay migration', *GLQ*, 2: 253–77.

Whatmore, S. (1990), *Farming Women: Gender, Work and Family Enterprise*, London: Macmillan.

Whatmore, S. (1991), 'Lifecycle or patriarchy? gender divisions in family farming', *Journal of Rural Studies*, 7: 71–6.

Whatmore, S. (1994), 'Theoretical achievements and challenges in European rural gender studies', in Van der Plas, L. and Fonte, M. (eds), *Rural Gender Studies in Europe*, Assen, The Netherlands: Van Gorcum.

Whatmore, S. (1999), 'Hybrid geographies: rethinking the "human" in human geography', in Massey, D., Allen, J. and Sarre, P. (eds), *Human Geography Today*, Cambridge: Polity Press.

Whatmore, S., Marsden, T. and Lowe, P. (eds) (1994), *Gender and Rurality*. Critical Perspectives on Rural Change Series V1, London: David Fulton.

Wheelock, J., Ljunggren, E. and Bains, S. (1999), 'Between the household and the market: a comparative study of entrepreneurs in Norway and England', paper presented at the Gender and Rural Transformations in Europe Conference, University of Wageningen, The Netherlands, October 1999.

Williams, R. (1973), *The Country and the City*, London: Chatto and Windus.

Williams, W. (1956), *The Sociology of an English Village: Gosforth*, Routledge and Kegan Paul: London.

Wilson, A. (1991), *The Culture of Nature: From Disney to the Exxon Valdez*. Oxford: Blackwell.

Women and Geography Study Group (1984), *Geography and Gender: An Introduction to Feminist Geography*, London: Hutchinson.

Women and Geography Study Group (1997), *Feminist Geographies: Explorations in Diversity and Difference*, London: Longman.

Woods, M. (1997), 'Discourses of power and rurality: local politics in Somerset in the 20th Century', *Political Geography*, 16: 453–78.

Woods, M. (1998), 'Advocating rurality? The repositioning of rural local government', *Journal of Rural Studies*, 14: 13–26.

Woodward, R. (1998), ' "It's a man's life!": soldiers, masculinity and the countryside', *Gender, Place and Culture*, 5: 277–300.

Wright, S. (1992), 'Image and analysis: new directions in community studies', in Short, B. (ed.), *The English Rural Community*, Cambridge: Cambridge University Press.

Young, I.M. (1990), 'The ideal of community and the politics of difference', in Nicholson, L. (ed.), *Feminism/Postmodernism*, London: Routledge.

Zelinsky, W., Monk, J. and Hanson, S. (1982), 'Women and Geography: a review and prospectus', *Progress in Human Geography*, 6: 317–66.

Index